Get the most from this book

Everyone has to decide his or her own revision strategy, but it is essential to review your work, learn it and test your understanding. These Revision Notes will help you to do that in a planned way, topic by topic. Use this book as the cornerstone of your revision and don't hesitate to write in it – personalise your notes and check your progress by ticking off each section as you revise.

Tick to track your progress

Use the revision planner on pages iv and v to plan your revision, topic by topic. Tick each box when you have:

- revised and understood a topic
- tested yourself
- practised the exam questions and gone online to check your answers and complete the quick quizzes.

You can also keep track of your revision by ticking off each topic heading in the book. You may find it helpful to add your own notes as you work through each topic.

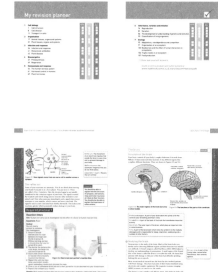

Features to help you succeed

Definitions and key words

Clear, concise definitions of essential key terms are provided where they first appear. Key words from the specification are highlighted in bold throughout the book.

Exam tips

Expert tips are given throughout the book to help you polish your exam technique in order to maximise your chances in the exam.

Now test yourself

These short, knowledge-based questions provide the first step in testing your learning. Answers are at the back of the book.

Summaries

The summaries provide a quick-check bullet list for each topic.

Exam practice

Practice exam questions are provided for each topic. Use them to consolidate your revision and practise your exam skills.

Online

Go online to check your answers to the exam questions and try out the extra quick quizzes at **www.hoddereducation.co.uk/ myrevisionnotesdownloads**

Revision activities

These activities will help you to understand each topic in an interactive way.

Typical mistakes

The author identifies the typical mistakes candidates make and explains how you can avoid them.

H Where this symbol appears, the text to the right of it relates to higher tier material.

D0493414

My revision planner

REVISED   TESTED   EXAM READY

REVISED   TESTED   EXAM READY

Exam practice answers and quick quizzes at
www.hoddereducation.co.uk myrevisionnotesdownloads

Countdown to my exams

6–8 weeks to go

- Start by looking at the specification – make sure you know exactly what material you need to revise and the style of the examination. Use the revision planner on pages iv and v to familiarise yourself with the topics.
- Organise your notes, making sure you have covered everything on the specification. The revision planner will help you to group your notes into topics.
- Work out a realistic revision plan that will allow you time for relaxation. Set aside days and times for all the subjects that you need to study, and stick to your timetable.
- Set yourself sensible targets. Break your revision down into focused sessions of around 40 minutes, divided by breaks. These Revision Notes organise the basic facts into short, memorable sections to make revising easier.

REVISED ☐

2–6 weeks to go

- Read through the relevant sections of this book and refer to the exam tips and key terms. Tick off the topics as you feel confident about them. Highlight those topics you find difficult and look at them again in detail.
- Test your understanding of each topic by working through the 'Now test yourself' questions in the book. Look up the answers at the back of the book.
- Make a note of any problem areas as you revise, and ask your teacher to go over these in class.
- Look at past papers. They are one of the best ways to revise and practise your exam skills. Write or prepare planned answers to the exam practice questions provided in this book. Check your answers online and try out the extra quick quizzes at **www.hoddereducation.co.uk/ myrevisionnotesdownloads**
- Try out different revision methods. For example, you can make notes using mind maps, spider diagrams or flash cards.
- Track your progress using the revision planner and give yourself a reward when you have achieved your target.

REVISED ☐

One week to go

- Try to fit in at least one more timed practice of an entire past paper and seek feedback from your teacher, comparing your work closely with the mark scheme.
- Check the revision planner to make sure you haven't missed out any topics. Brush up on any areas of difficulty by talking them over with a friend or getting help from your teacher.
- Attend any revision classes put on by your teacher. Remember, he or she is an expert at preparing people for examinations.

REVISED ☐

The day before the examination

- Flick through these Revision Notes for useful reminders, for example the exam tips and key terms.
- Check the time and place of your examination.
- Make sure you have everything you need – extra pens and pencils, tissues, a watch, bottled water, sweets.
- Allow some time to relax and have an early night to ensure you are fresh and alert for the examination.

REVISED ☐

My Exams

GCSE Biology Paper 1

Date:..

Time: ....

Location: ....

GCSE Biology Paper 2

Date:..

Time: ....

Location: ....

1 Cell biology

Cell structure

Eukaryotes and prokaryotes

Eukaryotes

Eukaryotic organisms, or eukaryotes, have cells with a nucleus. Animal and plant cells are eukaryotic. You are a eukaryote, so almost all your cells have a nucleus containing your **DNA** in their middle.

Prokaryotes (bacteria)

Prokaryotic organisms, or prokaryotes, have cells without a nucleus. Bacterial cells are prokaryotic, so bacteria are prokaryotes. All bacteria are single celled and are usually smaller than eukaryotic cells. Bacterial cells do not have a nucleus. Their chromosomal DNA is found within their cytoplasm. Prokaryotic bacterial cells also contain small rings of DNA called plasmids.

DNA (deoxyribonucleic acid): The genetic information found in all living organisms.

Ribosomes: Subcellular structures found in the cytoplasm of cells in which synthesis occurs.

Respiration: A chemical reaction that occurs in mitochondria found in the cytoplasm of cells which releases energy from glucose for life processes.

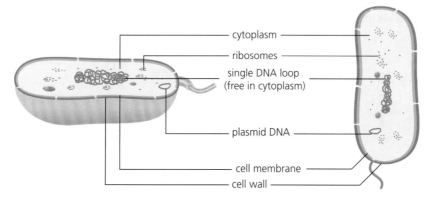

cytoplasm
ribosomes
single DNA loop (free in cytoplasm)
plasmid DNA
cell membrane
cell wall

Figure 1.1 A bacterial cell as three- and two-dimensional diagrams.

The components of prokaryotic cells and their functions are shown in Table 1.1.

Revision activity

Draw out this table with only the headings along the top and the first column on the left. Try to fill in the rest of the table from memory to help you to revise.

Table 1.1 The components of bacterial cells and their functions.

Component	Structure and function
Cytoplasm	This fluid is part of the cell inside the cell membrane. It is mainly water and it holds other components such as **ribosomes**. Here most of the chemical reactions in the cell happen (such as the making of proteins in ribosomes).
Cell wall	Like those of plants and fungi, bacterial cells have a cell wall to provide support. However, unlike plant cell walls this is not made of cellulose. The cell membrane is found on the inside surface of the cell wall.
Single DNA loop (DNA not in chromosomes)	DNA in prokaryotes is not arranged in complex chromosomes as in eukaryotic cells. It is not held within a nucleus.
Plasmids	These are small, circular sections of DNA. They provide genetic variation for bacteria.
Cell membrane	This controls what substances go in and out of a cell. It also has internal extensions that have enzymes attached to them. **Respiration** occurs in these enzymes.
Ribosome	Proteins are made by ribosomes, which are present in the cytoplasm.

Animal and plant cells

Generalised (typical) animal cells

A generalised animal cell is seen in Figure 1.2.

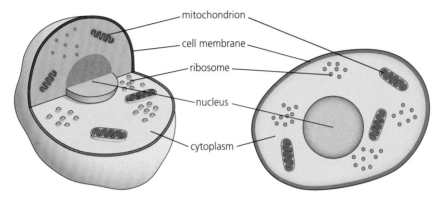

Figure 1.2 A generalised animal cell as three- and two-dimensional diagrams.

Generalised animal cells have components in common with bacterial cells described on the previous page. Both possess a cytoplasm in which most chemical reactions occur, and a cell membrane that controls what enters and exits the cell. The functions of their additional components are found in Table 1.2.

Generalised (typical) plant cells

A generalised plant cell is seen in Figure 1.3.

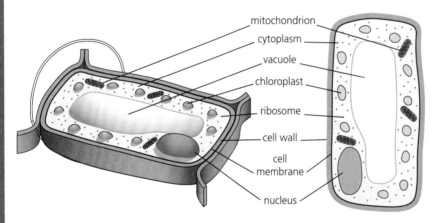

Figure 1.3 A generalised plant cell as three- and two-dimensional diagrams.

Generalised plant cells have all the components previously described for animal cells. Both possess a nucleus, cytoplasm, membrane, mitochondria and ribosomes. Additionally, many plant cells possess a vacuole, which is a store of sugary cell sap, chloroplasts where photosynthesis occurs and a cell wall to provide structure.

Revision activity

Draw out Table 1.2 with only the headings along the top and the first column on the left. Try to fill in the rest of the table from memory to help you to revise.

Exam tip

You should be able to explain how the structures in Table 1.2 are related to their functions.

Typical mistake

The nucleus of a cell is much larger than that of an atom (in chemistry). The nucleus of a cell is made up of thousands of atoms each with their own nucleus.

Organelle: A part of a cell with a specific function.

Mineral ions: Substances that are essential for healthy plant growth, e.g. nitrates and magnesium.

Table 1.2 The components present in generalised animal and plant cells and their functions.

Component	Present in:		Function
	Animal cells	Plant cells	
Nucleus	Yes	Yes	This contains the DNA or genetic information of an organism arranged into chromosomes.
Cytoplasm	Yes	Yes	This fluid is part of the cell inside the cell membrane, which is mainly made of water. It holds other components like mitochondria and ribosomes.
Cell membrane	Yes	Yes	This controls which substances go in and out of a cell.
Mitochondrion (plural mitochondria)	Yes	Yes	A small cell **organelle** in the cytoplasm in which respiration releases energy from glucose.
Ribosome	Yes	Yes	Proteins are made by ribosomes, which are present in the cytoplasm.
Vacuole	No	Yes	The vacuole is found in the middle of many plant cells and contains cell sap. Dissolved sugars and **mineral ions** are stored here.
Chloroplast	No	Yes	Photosynthesis occurs in chlorophyll found in chloroplasts.
Cell wall	No	Yes	Like bacteria and fungi, plant cells have a wall to provide support. This wall is made from cellulose.

Cell specialisation

You learned about generalised animal and plant cells on the previous page. But complex multicellular organisms like you, and other animals and plants, are not made up of just one type of cell. There are around 200 different types of cell in your body. Each cell type has become specialised to complete a specific function. Cells can develop specific components during this process.

Your cells, and those of many animals, became specialised, or differentiated, before you were born. Unlike animal cells, many plant cells retain their ability to **differentiate** throughout their entire life.

Sperm cell

Sperm cells possess a tail to propel them towards the **ovum** (egg). For their very small size, they have many **mitochondria** to release energy during respiration. Their nucleus contains the DNA from the father, which will make up half of the DNA of the new organism.

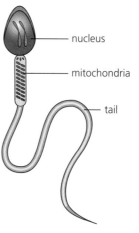

nucleus

mitochondria

tail

Figure 1.4 The parts of a sperm cell.

Differentiate: To develop into something different.

Ova (singular ovum): Eggs.

Mitochondrion: A small cell organelle, in which respiration occurs, found in the cytoplasm of eukaryotic cells.

Typical mistake

It is important that you can explain why a specialised cell is adapted, not just state its adaptation.

Exam tip

You should be able to explain the importance of cell differentiation.

Nerve cell

Nerve cells (neurones) pass electrical signals around your body to control and coordinate your actions. They possess a long **axon**, along which the electrical signals quickly pass. This is insulated by the **myelin sheath**. Nerve cells possess branching nerve endings that can communicate with surrounding cells.

> **Axon**: The extension of a nerve cell along which electrical impulses travel.
>
> **Myelin sheath**: The insulating cover along an axon that speeds up the electrical signal.

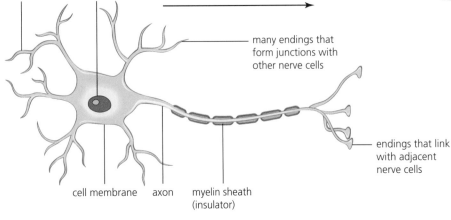

Figure 1.5 **The parts of a nerve cell.**

Muscle cells

Muscle cells contract and relax to move parts of our bodies. This movement can be automatic or involuntary like your heartbeat, or voluntary like moving your fingers to type an email. They possess large numbers of mitochondria to release the energy from glucose in respiration.

Root hair cell

Root hair cells possess a long extension into the soil. This extension massively increases the surface area of the cell, meaning that it can absorb more water and minerals from the soil.

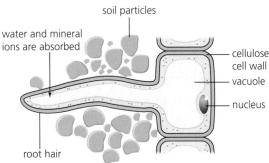
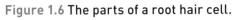

Figure 1.6 **The parts of a root hair cell.**

Xylem cell

Xylem cells form long tubes that run from the roots to the leaves of plants. They carry water to the leaves for photosynthesis. This process is called transpiration. Xylem cells are dead and have eroded-away ends to allow the water to move more easily.

Phloem cell

Phloem cells carry dissolved glucose made during photosynthesis from the leaves to all other parts of the plant. This process is called **translocation**. Phloem cells are alive. They have specialised endings to their cells called sieve plates, which allow water to flow more easily through them. They have companion cells that support the **metabolism** of phloem cells.

> **Phloem**: Living cells that carry sugars made in photosynthesis to all cells of a plant.
>
> **Translocation**: The movements of sugars made in photosynthesis from the leaves of the plant.
>
> **Metabolism**: The sum of all the chemical reactions that happen in an organism.

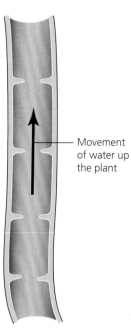

Movement of water up the plant

Figure 1.7 The parts of a xylem tube.

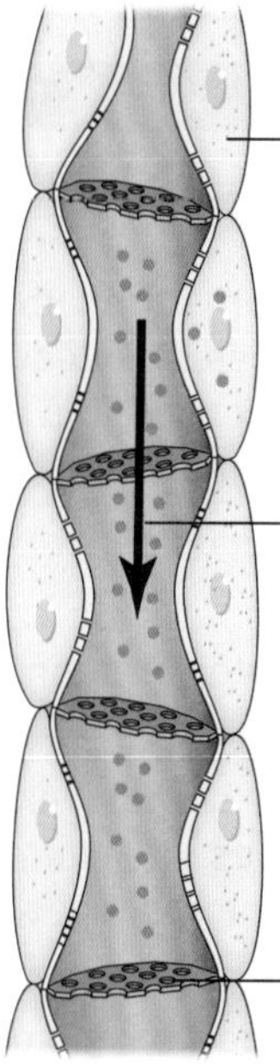

companion cell
translocation of sucrose
sieve plate

Figure 1.8 The parts of a phloem tube.

Exam tip

You should be able to explain how the structure of specialised cells relates to their function.

Revision activity

You could make flash cards with a diagram of the specialised cell on one side and its adaptation on the other.

Microscopy

REVISED

Light microscopes

Microscopes allow you to see structures too small to view with your eyes alone. Light microscopes were the first type to be developed, perhaps by Dutch eye glass makers in the 1590s. The design of light microscopes has developed since then but the basic principle of using glass lenses to magnify images is the same. A typical light microscope is shown in Figure 1.9 and the functions of its components are given in Table 1.3.

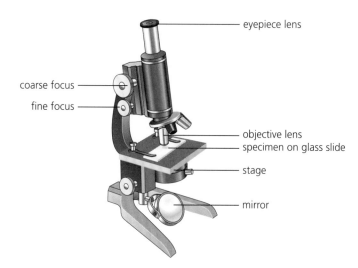

eyepiece lens
coarse focus
fine focus
objective lens
specimen on glass slide
stage
mirror

Figure 1.9 A labelled diagram of a light microscope.

Table 1.3 The functions of the parts of a light microscope.

Part	Function
Eyepiece lens	You look through this lens to see your sample. This is often ×10.
Objective lens	Usually there are three to choose from (often ×5, ×10 and ×25). The smallest will be the easiest to focus, so select this first. When you have focused this lens try a different one with a greater magnification.
Stage	This holds the sample securely, often using two metal clips.
Specimen	This is usually placed in a drop of water or stain on a microscope slide under a very thin glass cover slip.
Mirror	This reflects the light up through the sample, and then the objective and eyepiece lenses into your eyes. In more expensive/advanced microscopes the mirror is replaced by a light source.
Course focus	This quickly and easily moves the stage up and down to focus on the sample.
Fine focus	This sensitively and slowly moves the stage up and down to allow you to make your image very sharp.

The total magnification of a light microscope is calculated by this formula:

total magnification = magnification of eyepiece lens × magnification of objective lens

Required practical 1

Light microscopy

Aim: To use a light microscope to observe, draw and label a selection of plant and animal cells.

Equipment: Light microscope, slides with samples, drawing equipment.

Method:
1 Place your slide onto the stage of the microscope.
2 Observe using the lowest power objective lens.
3 Use the course and then fine focus dials to focus the image.
4 Change to the next highest objective lens and refocus using the fine focus.
5 Repeat with higher objective lenses if appropriate.
6 Draw scientific images of your observations.
7 Record the magnification that you used on your images.

Electron microscopes

Light microscopes have a maximum magnification of around 1000 times. This means they can be used to see cellular components like mitochondria and ribosomes.

However, in the 1930s German scientists developed the **electron microscope**. This microscope uses beams of **electrons** in place of light to magnify an image. The wavelength of the electrons can be a 100 000 times smaller than light. This fact allows electron microscopes

> **Revision activity**
>
> Draw out this table with only the headings along the top and the first column on the left. Try to fill in the rest of the table from memory to help you to revise.

> **Exam tip**
>
> You should be able to show that you can estimate the scale and size of cells.

> **Electron microscope**: A microscope that uses electron beams in place of light to give a higher magnification.
>
> **Electron**: A negatively charged, tiny subatomic particle that is found in shells surrounding the nucleus of the atom.

to take images at higher magnifications. This first electron microscope was an example of a transmission electron microscope, which takes two-dimensional high magnification images.

The design of electron microscopes developed during the 1930s. A second type of microscope called a scanning electron microscope was also developed by German scientists. This takes three-dimensional high magnification images.

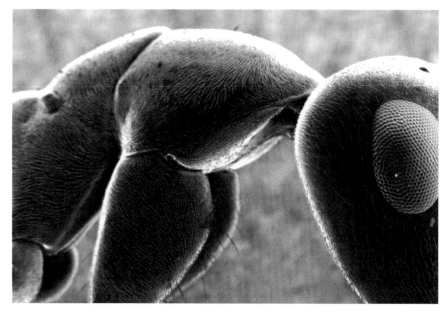

Figure 1.10 **A three-dimensional image of an ant taken with a scanning electron microscope.**

> **Exam tip**
>
> You should be able to explain how microscopes have developed and how electron microscopes have helped us understand the sub-cellular structure of cells.

Magnification calculations

REVISED

The magnification of an image can be calculated by the following formula:

$$\text{magnification} = \frac{\text{size of image}}{\text{size of real object}}$$

> **Exam tip**
>
> You should be able to express your answers in standard form, if appropriate.

Culturing micro-organisms

REVISED

Bacteria copy themselves in a process similar to **mitosis** called binary fusion.

Required practical 2

Growing bacteria

Aim: To investigate the effect of **antiseptics** or **antibiotics** on bacterial growth using agar plates and measuring zones of inhibition.

Equipment: Agar Petri dishes covered with bacteria, paper discs, three antibiotic or antiseptic solutions, distilled water, tweezers.

Method:
1 Draw lines on the bottom of a Petri dish to split it into four sections.
2 Dip a disc of paper in an antibiotic or antiseptic solution and place on top of the bacteria in the middle of one of the four sections.
3 Repeat with two more discs using different antibiotics.

> **Mitosis:** Cell replication that produces two identical copies of a diploid cell.
>
> **Antiseptic:** A substance applied to the skin or another surface to destroy pathogens.
>
> **Antibiotics:** A group of medicines that kills bacteria but not viruses. The first antibiotic was discovered by Sir Alexander Fleming.

4 Dip the final disc into distilled water, as a control, and place in the remaining section.

5 Incubate for several days.

Aseptic technique: This is a series of techniques used when studying micro-organisms to stop contamination. This technique includes sterilisation of equipment in alcohol or a flame, wearing gloves or other protective equipment and cleaning before and afterwards using antiseptics.

Expected results: If the bacteria have been killed by the antibiotic or antiseptic, then an area of clearing will be seen around the disc. The bigger the clearing the more effective the antibiotic or antiseptic. It is likely no clearing will be seen around the control disc.

Now test yourself

TESTED ☐

1 Define the term DNA.
2 Explain the difference between prokaryotes and eukaryotes.
3 What are plasmids?
4 State the cell components present in plant cells that are not found in animal cells.
5 What process happens in mitochondria?
6 What is the name given to the part of a neurone along which electrical impulses travel?
7 Why do plant roots have root hair cells?
8 Which plant vessels are involved in transpiration?
9 How do we calculate the magnification of a light microscope?
10 Of what are both sterilising equipment in a flame and wearing gloves both examples?

Answers on p. 113

> **Exam tip**
>
> You should be able to describe how to prepare an uncontaminated culture using aseptic technique.

> **Exam tip**
>
> You should be able to calculate the area of colonies using the formula πr^2. You should also be able to calculate the number of bacteria in a population after a certain time if you are given the starting number and how often they divide.

> **(H) Exam tip**
>
> You should be able to express your answers in standard form.

Cell division

Chromosomes

REVISED ☐

Almost all your cells contain a nucleus with one copy of your DNA in. This is your **genome**. Unless you are an identical twin, your genome is unique. It is highly unlikely that anyone has or will ever be born with the same genome as you. Your genome is made up of sections called genes that contain the DNA code to make proteins. Other sections do not make proteins. Currently we are not sure of the purpose of these non-coding regions so we call them 'junk' DNA.

Your genome is made from about 2 metres of DNA. This is too much to be simply arranged loosely in the cytoplasm like prokaryotic bacterial cells (see page 1). It is arranged into 23 pairs of smaller sections called chromosomes. There are 46 in total, but we often say 23 pairs instead. This convention reminds us that chromosomes come in pairs; 23 from our mum in her ovum and 23 from our dad in his sperm. Cells with an entire copy of a genome in them are called **diploid**.

Chromosomes are long thin structures, made from coiled up DNA. They taper in the middle.

Sperm and ova are called sex cells, or **gametes**, and have half the DNA of normal body cells. They are called **haploid**. Two haploid gametes

> **Genome:** One copy of all the DNA found in your diploid body cells.
>
> **Diploid:** Describes a cell or nucleus of a cell that has a paired set of chromosomes.
>
> **Gametes:** Sex cells, e.g. sperm, ova, spores and pollen.
>
> **Haploid:** Describes a cell or nucleus of a gamete that has an unpaired set of chromosomes.

join during fertilisation to make a diploid body cell. You inherit two copies of almost all genes; one from each parent. These copies are called **alleles**.

Humans have 23 pairs of chromosomes, whilst many other animals and plants have different numbers. Lettuces have nine pairs and mosquitos have only have one pair; or one chromosome from each parent.

> **Alleles:** Two copies of the same gene, one from your mother and the other from your father.

Mitosis and the cell cycle

REVISED

Your diploid body cells are continually dying and need replacing. It is unlikely that you have any of the blood cells still alive in your body from a year ago. They have all been replaced at least once since then. Mitosis and the cell cycle explains how this process occurs.

Without mitosis, your body would not have grown from the one diploid cell formed when your father's sperm fertilised your mother's ovum. Additionally, any cuts, burns or other damage to your body would not have healed. Mitosis is essential for growth, development and repair of multicellular organisms.

Mitosis exchanges damaged or old cells with identical replacements. Mistakes in this process could lead to cancer.

During the cell cycle, all chromosomes are copied. This doubles the number of chromosomes from 23 pairs to 46 pairs, or 92 in total. Then the number of cellular components like ribosomes and mitochondria doubles. The cell finally splits into two identical 'daughter' cells, each with an entire copy of the organism's genome.

Steps in mitosis

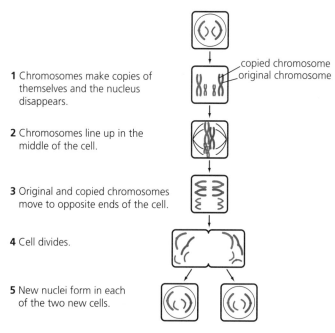

1 Chromosomes make copies of themselves and the nucleus disappears.

copied chromosome
original chromosome

2 Chromosomes line up in the middle of the cell.

3 Original and copied chromosomes move to opposite ends of the cell.

4 Cell divides.

5 New nuclei form in each of the two new cells.

Figure 1.11 The main steps in mitosis for a cell with just two pairs of chromosomes. Note that the two new cells at the bottom of the diagram are identical to each other and also the original 'parent' cell at the top in terms of chromosome number.

> **Exam tip**
>
> You should be able to describe the cell cycle but not the individual stages of mitosis. You should be able to recognise situations in which mitosis is occurring.

Stem cells possess the ability to develop into other types of cell.

Stem cells in mammals

There are two types of stem cell found in you, and other mammals. During fertilisation, your mother's ovum and father's sperm fused. This formed one diploid embryonic stem cell. This divided by mitosis until you were formed. For the first 9 weeks of your life your cells were not specialised. They had not differentiated. They remained embryonic stem cells. These can develop into any of the 200 cell types you possess. These are called **totipotent** cells.

Adult stem cells are the second type of stem cell found in your body. Confusingly, you started to develop adult stem cells much earlier than you became or will become an adult. Adult stem cells are found in specific locations such as the bone marrow and the nose. Here they can only develop into one or two cell types. Adult stem cells in your bone marrow can develop into blood cells, whilst those in your nose can develop into nerve cells. These are called **multipotent** cells.

Other animals like lizards can shed their tails, if caught by them, to avoid being killed by a predator. They are then able to regrow their tail from their stem cells. If one leg of a starfish is severed, it will grow four new ones, and the original starfish will grow one one new leg.

Stem cells and differentiation in plants

Plant stem cells are found in specific locations called **meristems**. These regions are in the tips of shoots and roots. Much of a plant's growth occurs in these regions. Unlike your adult stem cells, plant stem cells retain the ability to differentiate throughout their life. This fact means we can take a cutting of a small plant stem and place it in soil. Here stem cells will start to develop into roots and a cloned copy will have been formed.

Stem cell research

Stem cell research uses stem cells to develop future medical treatments that could treat paralysed patients by making new nerve cells to transplant into a damaged spinal cord, or replace injured or non-working organs such as the pancreases of **diabetes** patients. Totipotent embryonic stem cells are more useful in research.

Using a person's own stem cells in medical treatments means their bodies are far less likely to reject them like a transplant from another person. The process of making an embryo with the same genes as the parent for this reason is called therapeutic cloning. However, there is a small possibility of transferring virus infections in this process.

Stem cells research is an **ethical issue**. This means some people disagree with it for religious or moral reasons. Many people donate unused fertilised ova from *in-vitro* fertilisation for stem cell research. Controversy surrounds the use of these cells. Are they alive or a life? Because of these issues, tight regulations surround all scientific studies.

Stem cell: An undifferentiated cell that can develop into one or more types of specialised cell.

Totipotent: Describes a stem cell that can develop into any type of specialised cell.

Multipotent: Describes a stem cell that can only develop into several types of specialised cell.

Meristem: An area of a plant in which rapid cell division occurs, normally in the tip of a root or shoot.

Diabetes: A non-communicable disease that reduces control of blood glucose concentrations.

Ethical issue: An idea (issue) some people disagree with for religious or moral reasons.

In-vitro **fertilisation (IVF)**: A medical procedure in which ova are fertilised outside of a woman, then placed into her uterus to develop into a baby.

Exam tip

You should be able to describe the function of stem cells in embryos, adult animals and plant meristems.

Now test yourself

TESTED

11 Define the term genome.
12 What are chromosomes?
13 State the only two haploid body cells.
14 What are the products of mitosis?
15 What could mistakes in the process of mitosis lead to?
16 What is different about adult and embryonic stem cells?
17 Define the term totipotent.
18 What name is given to an area of a plant in which rapid cell division occurs?
19 In what regions of plant are its meristems?
20 What is the difference between stem cells in plants and animals?

Answers on p. 113

Transport in cells

Diffusion

REVISED

Diffusion is the spreading out of particles resulting in their **net** movement from an area of high to lower concentration. This happens naturally and does not require energy, so we call it a passive process. Because diffusion always happens from high to lower concentrations, we say it occurs down a **concentration gradient**.

Particles of gases and liquids can diffuse. Those of solids have fixed positions and so cannot.

Examples of diffusion

Diffusion occurs in many places in your body. You breathe oxygen into the **alveoli** in your lungs. Here oxygen diffuses from a high concentration in your alveoli to a lower concentration in your red blood cells. When these cells absorb oxygen, they turn from a low to high concentration. They then move through your blood vessels to your body's cells. Because these cells have been respiring they have been using oxygen. This means they have a low concentration of oxygen. So, oxygen moves by diffusion from the high concentration in your red blood cells to the lower concentration in your body's cells.

This process is repeated in reverse with carbon dioxide. Carbon dioxide is produced during respiration by your body's cells. It dissolves straight into your **blood plasma**, and is not absorbed by red blood cells, as with oxygen.

Sugars such as glucose are produced by your digestive system when

1

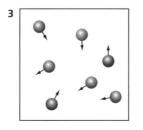

2

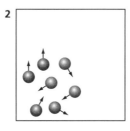

3

Figure 1.12 **Molecules in a gas spread out by diffusion.**

> **Diffusion:** The net movement of particles from an area of high concentration to an area of lower concentration.
>
> **Net:** Overall.
>
> **Concentration gradient:** A measurement of how a concentration of a substance changes from one place to another.
>
> **Alveoli:** A tiny air sac found in the lungs through which gases exchange between blood and air.
>
> **Blood plasma:** The straw-coloured liquid that carries our blood cells and dissolved molecules.

carbohydrates are broken down. These therefore exist at high concentration in your small intestine. So, glucose moves by diffusion from your small intestine through **villi** to the lower glucose concentration in your blood. When your blood absorbs glucose, it turns from a low to high concentration. Blood moves round your body to your body's cells. Because these cells have been respiring they have been using glucose. This means that the glucose now moves by diffusion from the higher concentration in your blood to the lower one in your cells.

Some of your cells make **urea** as a waste product. This diffuses from a higher concentration in the cells to a lower one in the blood. Urea is transported to the kidney where it is **excreted**.

> **Villi (singular villus):** Tiny finger-like projections that increase the surface area of the small intestine.
>
> **Urea:** A key waste product of protein metabolism in mammals that is excreted in urine.
>
> **Excretion:** The removal of substances from cells or organisms.

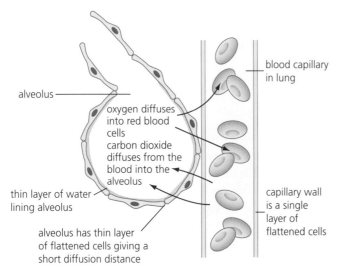

Figure 1.13 Diffusion of gases between an alveolus and a blood capillary in the lung.

Diffusion in other organisms

Your lungs are extremely effective at absorbing oxygen. But not all animals have lungs. Insects have small tubes that run into their bodies, into which gases diffuse. This system is much less sophisticated than your lungs, and so insects cannot complete as much gas exchange as humans. They do not possess efficient exchange surfaces like your lungs or the gills of fish. This means insects are always limited by size. Their maximum size is in part determined by the distance that oxygen can easily diffuse into their cells. In fish gills, water passes in the opposite direction to the blood meaning that maximum diffusion of oxygen into their blood occurs.

Smaller organisms like insects or single-celled organisms have a larger surface area to volume ratio than larger organisms. This means they can survive without specialised exchange surfaces.

Factors that affect diffusion

The factors that affect diffusion are shown in Table 1.4.

> **Exam tip**
>
> You should be able to describe how the factors in Table 1.4 affect diffusion.

> **Exam tip**
>
> You should be able to explain why exchange surfaces and a transport system are needed. In particular, you should be able to explain how the small intestine and lungs in mammals, gills in fish and the roots and leaves in plants are adapted.

Table 1.4 The factors that affect diffusion.

Factor	How
Difference in concentrations	If two concentrations are similar the rate of diffusion will occur slowly. The larger the difference in concentration, the quicker the rate of diffusion.
Temperature	At higher temperatures, all particles have more **kinetic energy**. The higher the temperature, the quicker the rate of diffusion.
Surface area	More diffusion can occur over a large surface area. The larger the surface area, the quicker the rate of diffusion.

Kinetic energy: The store of movement energy.

Revision activity

Draw out this table with only the headings along the top and the first column on the left. Try to fill in the rest of the table from memory to help you to revise.

Osmosis

REVISED

Osmosis is the spreading out of water particles resulting in the net movement of water from an area of high to lower water concentration across a **partially permeable** membrane. So, osmosis is the diffusion of water across a membrane.

This happens naturally and does not require energy, so we call it a passive process. Because osmosis always happens from a high to a lower concentration, we say it occurs down a concentration gradient.

Osmosis: The net diffusion of water from an area of high water concentration to an area of lower water concentration across a partially permeable membrane.

Partially permeable: Allowing only substances of a certain size through.

Examples of osmosis

When it rains, the soil becomes wet. It has a high concentration of water particles. This concentration is often higher than that in the plant's root hair cells, so water moves by osmosis (or osmoses) into the plant. It moves from a high to a lower concentration across the membrane of the root hair cells.

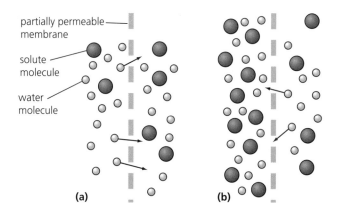

Figure 1.14 Two partially permeable membranes. Both membranes have the same concentration of solution on the right-hand side. However, water moves in opposite directions through the membranes.

We can see osmosis when we put cells into different solutions. Your red blood cells and the plasma they are surrounded by are the same concentration. So, water can move from the cells into the plasma and in reverse, but no net movement occurs. Solutions with the same overall concentration are called isotonic.

If a red blood cell is put into a solution of salty water it will shrivel and shrink. There is a higher concentration of water in the cell than in the surrounding solution. So, the water will move by osmosis from the cell into the solution. If one solution has a higher concentration than another one, we call the first one **hypertonic**.

If a red blood cell is put into a solution of distilled water it will swell and could burst. There is a lower concentration of water in the cell than in the surrounding solution. So, the water will move by osmosis from the solution into the cell. If one solution has a lower concentration than another one, we call the first one **hypotonic**.

  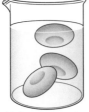

correct concentration   low concentration   high concentration
of water   of water (brine)   of water

Figure 1.15 This is what happens to red blood cells in solutions with different concentrations of water. (Not to scale.)

> **Hypertonic**: A solution with a higher concentration of solutes.
>
> **Hypotonic**: A solution with a lower concentration of solutes.

Required practical 3

Investigating osmosis

Aim: To investigate the effect of a range of concentrations of salt or sugar solutions on the mass of plant tissue.

Equipment: Boiling tubes, five salt or sugar solutions, distilled water, potato discs.

Method:
1 Cut six small discs of potato and individually record the mass of each.
2 Place one disc in a boiling tube of each of the five different salt or sugar solutions.
3 Place one disc in a boiling tube of distilled water to act as a control.
4 Wait for 30 minutes.
5 Remove each disc and re-measure its mass.
6 Calculate the percentage change in mass for each disc.

Results:
Plot a graph of concentration of salt or sugar solution and change in mass. If the solution contains more sugar or salt than the potato disc, then the percentage change in mass will be lower than 100% because water moves from a high concentration of water in the disc to a lower concentration in the solution.

If the solution contains less sugar or salt than the potato disc, then the percentage change in mass will be greater than 100% because water moves from the high water concentration in the solution to the lower concentration in the disc.

If the disc stays the same mass, then the concentrations are the same and no net movement has occurred.

> **Exam tip**
>
> You should be able to calculate the rate at which water is absorbed into or lost from a plant, and calculate the overall percentage gain or loss of mass.

Active transport

Active transport is the net movement of particles from an area of lower to higher concentration. So, active transport reverses the effects of diffusion.

This movement does not happen naturally and so requires the use of energy. It is therefore not a passive process like diffusion and osmosis, but an active process. Because active transport always happens from a lower to a higher concentration, we say that it occurs up a concentration gradient.

> **Active transport**: The net movement of particles from an area of lower concentration to an area of higher concentration using energy.

Examples of active transport

Water moves from a high concentration in the soil to a lower concentration in the plant by osmosis. This movement does not require energy. But what happens to mineral ions found in the soil that plants need to absorb? Minerals are present at higher concentrations in the plant than in the soil. So, they cannot move into the plant by diffusion. Plants need to absorb these mineral ions from the lower concentration in the soil to the higher concentration in the plant. This movement requires active transport, which uses energy.

This process also occurs in your digestive system. When you have just digested a meal, glucose is found at a high concentration inside your small intestine. It moves by diffusion into your blood. This does not require energy. But what happens when most of this glucose has been absorbed? The glucose is now at a higher concentration in your blood and so cannot be absorbed by diffusion. Your body absorbs the last of the glucose into your blood by active transport. This final movement uses energy.

> **Exam tip**
>
> You should be able to describe how substances are transported into and out of cells by diffusion, osmosis and active transport. In doing so, you should be able to explain the differences between the three processes.

> **Revision activity**
>
> Draw a table with the headings diffusion, osmosis and active transport. In the table, describe each process (perhaps with a diagram) and give examples of where each occurs.

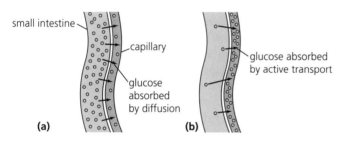

Figure 1.16 Look carefully at the glucose concentrations in the intestine and in the blood: a) diffusion, and b) active transport.

Now test yourself

21 Define the term diffusion.
22 In which two places does oxygen diffuse in the body?
23 Define the term osmosis.
24 How is osmosis different from diffusion?
25 What term do we give to a solution with a lower concentration of solutes?
26 By what process does water move into plant root hair cells?
27 Define the term active transport.
28 What does active transport require that diffusion does not?
29 What two processes help move glucose from the small intestine into the capillaries?
30 Why do plants absorb mineral ions from the soil by active transport?

Answers on p. 113

Summary

- Animal and plant cells are eukaryotic, so possess a nucleus. Bacterial cells are prokaryotic and so do not.
- Animal cells have a nucleus, cytoplasm, cell membrane, mitochondria and ribosomes. Plant cells have the same components as animal cells, as well as chloroplasts and a permanent vacuole filled with cell sap. Plant and algal cells have a cell wall made from cellulose.
- Bacterial cells are smaller than plant or animal cells. They have a cytoplasm, cell membrane and wall. Their DNA is found as a single loop. They may also have DNA plasmids.
- Cell specialisation allows cells to complete a specific function. Examples in animals include sperm, nerve and muscle cells. Examples in plants include xylem, phloem and root hair cells.
- As organisms develop their cells differentiate to become specialised. These cells develop different components to fulfil their function. Differentiation happens early in the life of most animals. Many plant cells can differentiate throughout their life.
- Microscope technology has developed over time. Electron microscopes have a much higher magnification and resolution than light microscopes. Magnification is calculated as the size of the image divided by the size of the real object.
- Bacteria divide by binary fission. They can be grown on agar plates or in nutrient broth. Aseptic techniques ensure bacterial cultures are not contaminated.
- The nucleus of eukaryotic cells contains chromosomes made from DNA. Chromosomes contain large numbers of genes. In diploid cells, chromosomes exist as pairs.

- The life of a cell is described in the cell cycle. Mitosis is a part of this cycle. During mitosis, chromosomes double before the cell divides into two identical daughter cells. Mitosis is important in the growth and development of organisms.
- Stem cells are undifferentiated. In humans, embryonic stem cells can develop into many different cells types, whilst adult stem cells can only develop into a much smaller number of cell types. Treatment with stem cells, including therapeutic cloning, could help diabetic or paralysed patients. Stem cell research is an ethical issue.
- Plant stem cells are found in regions called meristems and can develop into any cell type. They can be used in tissue culture to produce clones of rare plants or of important crops.
- Diffusion is the net movement of particles from an area of high to lower concentration. Oxygen and carbon dioxide move in and out of cells by diffusion. The rate of diffusion is affected by the concentration gradient, temperature and surface area.
- Larger organisms require exchange surfaces such as lungs or gills to maximise diffusion.
- Osmosis is the net movement of water from a high to a lower concentration across a membrane. Water moves into roots and through plants by osmosis.
- Active transport is the net movement of particles from an area of low to higher concentration and requires energy. Plants absorb mineral ions by active transport. You absorb the last glucose molecules from your small intestine into your blood by active transport.

Exam practice

1 What is present in animal and plant cells but not bacterial ones? [1]
   A Cytoplasm   C Nucleus
   B Cell membrane   D Cellulose cell wall
2 Which type of cell is prokaryotic? [1]
   A Plant   C Bacterial
   B Animal   D Fungal
3 Define mitosis. [1]
4 How do human embryonic and adult stem cells differ? [2]
5 Describe the differences between light and electron microscopes. [2]
6 What is the total magnification of a light microscope with an eyepiece lens of ×10 and an objective lens of ×25? [1]
7 Describe how you would make a light microscope slide of a human cheek cell. [4]
8 Compare and contrast the processes of diffusion and active transport. [6]
9 Compare and contrast plant and animal cells. [6]

Answers and quick quizzes online

ONLINE

2 Organisation

Animal tissues, organs and systems

Levels of organisation in living organisms

Larger multicellular animals have several levels of organisation. From smallest to largest, they are:
1 Cells are the basic building blocks of all life.
2 Tissues are groups of cells with a similar structure and function.
3 Organs are groups of tissues that perform a specific function.
4 Organ systems are groups of organs with similar functions.
5 Organisms are made from organ systems.

Table 2.1 **Examples of levels of organisation.**

Organisational level	Examples
Cell	Nerve cell, muscle cell
Tissue	Nervous tissue, skin
Organ	Brain, heart
Organ system	Nervous system, digestive system
Organism	Human, frog

Revision activity

Draw out this table with only the headings along the top and the first column on the left. Try to fill in the rest of the table from memory to help you to revise.

The human digestive system

Your digestive system is about nine metres long and runs from your mouth to your anus.

It breaks down the large, **insoluble** bits of food that you eat. These are broken down into smaller, **soluble** pieces that can be absorbed into your blood. Once this happens they are transported around your body to the cells that need them.

Insoluble: Cannot dissolve.

Soluble: Can dissolve.

Functions of the parts of the digestive system

The locations of the parts of your digestive system are shown in Figure 2.1. The functions of these components are found in Table 2.2.

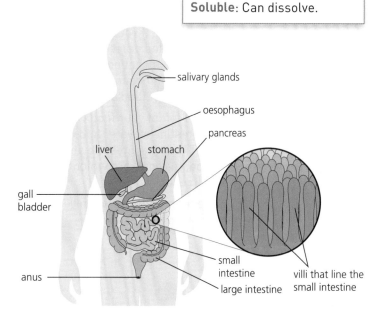

Figure 2.1 **The digestive system showing the location of the villi.**

Table 2.2 The parts of the digestive system and their functions.

Component	Function
Salivary glands	Salivary glands in your cheeks produce saliva. This saliva lubricates food as it passes along your oesophagus. Saliva also contains a carbohydrase **enzyme** called amylase that begins the break-down of carbohydrates into sugars.
Oesophagus	This short tube connects your mouth and stomach.
Stomach	Your stomach is a small bag about the size of your fist. It has ridges that allow it to increase in size when you eat food. Food is mixed with stomach acid to kill any **pathogens**. Stomach acid does not break down food. Protease enzymes are mixed with food to begin the break-down of proteins.
Liver	Food does not pass though the liver. The liver produces **bile**, which breaks down fats into smaller sections. This process is called emulsification. It increases the surface area of the food to allow lipase enzymes to work more effectively.
Gall bladder	Food does not pass through the gall bladder. Bile is stored in the gall bladder before being released into the small intestine.
Pancreas	Food does not pass through the pancreas. It produces carbohydrase, protease and lipase enzymes and releases these into the small intestine.
Small intestine	Digested food is absorbed into the blood in the small intestine. It is about six metres long. The surface of the small intestine is not smooth. It possesses millions of tiny finger-like projections called villi. These villi increase the surface area of the small intestine to allow more nutrients to be absorbed into the blood. Food is pushed through your small intestine by a process called **peristalsis**. This process is the rhythmical contraction and relaxation of muscles in the lining of the small intestine. This movement forces lumps of food along it.
Large intestine	All that is left of your food when it leaves the small intestine is water and fibre that you cannot digest. The large intestine absorbs water from this food, leaving fibre which forms your solid waste (faeces).
Anus	This opening controls when you release faeces when you go to the toilet.

Enzyme: A biological molecule that speeds up a chemical reaction.

Pathogen: A disease-causing micro-organism (bacterium, fungus or virus).

Bile: A green-coloured liquid produced by your liver, stored by your gall bladder and released into your small intestine to break down fats.

Peristalsis: The rhythmical contraction of muscle behind food in your digestive system to push it along.

Revision activity

Draw out this table with only the headings along the top and the first column on the left. Try to fill in the rest of the table from memory to help you to revise.

Revision activity

Pushing a tennis ball through a pair of tights is a good model for peristalsis pushing a bolus of food along your digestive system.

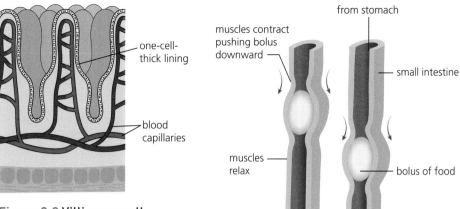

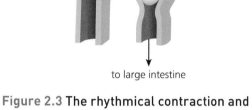

Figure 2.2 Villi are small, hair-like structures in your small intestine. Villi increase the surface area over which molecules of digested food are absorbed.

Figure 2.3 The rhythmical contraction and relaxation of the muscles that line much of the digestive system is called peristalsis.

Required practical 4

Food tests

Aim: To use qualitative reagents to test for a range of carbohydrates, **lipids** and proteins.

Equipment: Spotting tile, boiling tube, iodine solution, distilled water, Benedict's solution, water bath, Biuret solution, bung.

Method and results – starch test:
1 Place a small amount of food onto a spotting tile.
2 Add two drops of iodine solution.
3 If the food turns blue or black, starch is present.
  If it remains brown (the colour of iodine solution) then no starch is present.

Method and results – glucose test:
1 Place a small amount of food in a boiling tube.
2 Add 10 cm³ of distilled water.
3 Add 10 drops of Biuret solution to the boiling tube.
4 Heat in a water bath at 80 °C for 10 minutes.
5 If the solution turns orange or green, glucose is present.
  If it remains blue (the colour of Benedict's solution) then no glucose is present.

Method and results – protein test:
1 Place a small amount of food in a boiling tube.
2 Add 10 cm³ of distilled water.
3 Add 10 drops of Biuret solution to the boiling tube.
4 If the solution turns a light lilac colour, then protein is present.
  If it remains blue (the colour of Biuret solution) then no protein is present.

Method and results – oils test:
1 Place a small amount of food in a boiling tube.
2 Add 10 cm³ of distilled water.
3 Place bung in boiling tube and shake vigorously.
4 If an oil is present an emulsion will form and the water will turn cloudy.

Lipids: Fats or oils, which are insoluble in water.

Enzymes

Enzymes are biological catalysts. They speed up reactions and are not used up in them. This section focuses on the enzymes present in your digestive system. These enzymes break down large molecules of food into smaller ones. They are called break-down enzymes. There are other enzymes, however, that do the reverse. They join smaller molecules together to make larger ones. The enzyme involved in protein synthesis does this, for example. These are called synthesis enzymes.

> **Exam tip**
>
> You should be able to relate your knowledge of enzymes to metabolism.

Human digestive enzymes

There are three types of digestive enzyme. The molecules of food that they break down are called **substrates**. The three types of enzymes, their substrates and products and where they are found are shown in Table 2.3.

> **Substrate**: The molecule on which an enzyme acts.

Table 2.3 **The enzymes, substrates and products of the digestive system.**

Enzyme	Substrate	Product	Location
Carbohydrase	Carbohydrates	Sugars	Mouth, pancreas and small intestine
Protease	Proteins	Amino acids	Stomach, pancreas and small intestine
Lipase	Fats and oils (lipids)	Fatty acids and glycerol	Pancreas and small intestine

> **Revision activity**
>
> Draw out this table with only the headings along the top and the first column on the left. Try to fill in the rest of the table from memory to help you to revise.

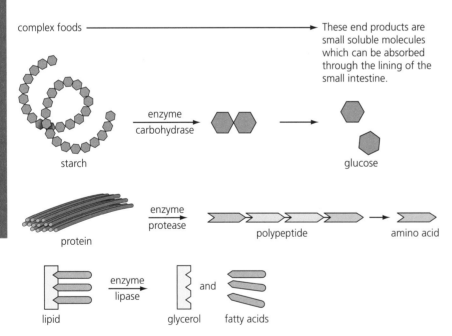

Figure 2.4 **The break-down of complex food molecules into small, soluble molecules that can be used.**

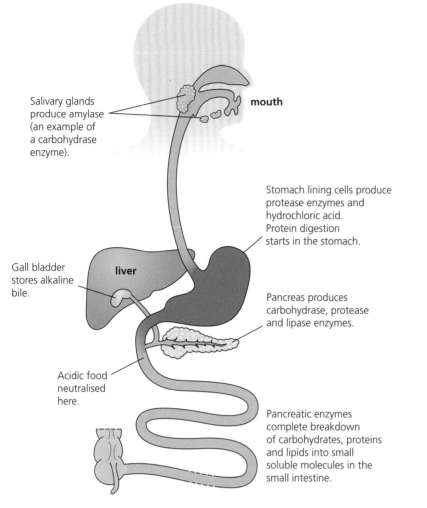

Salivary glands produce amylase (an example of a carbohydrase enzyme).

mouth

Gall bladder stores alkaline bile.

liver

Stomach lining cells produce protease enzymes and hydrochloric acid. Protein digestion starts in the stomach.

Pancreas produces carbohydrase, protease and lipase enzymes.

Acidic food neutralised here.

Pancreatic enzymes complete breakdown of carbohydrates, proteins and lipids into small soluble molecules in the small intestine.

Figure 2.5 Digestive enzymes control reactions that take place in the digestive system. No enzymes are made or used in the oesophagus, liver (bile is not an enzyme), gall bladder, large intestine or anus.

Bile

Bile is not an enzyme. It does not break down lipids into fatty acids and glycerol as lipase enzymes do. Bile is an emulsifier. It breaks down large globules of fat into smaller ones. This process increases the surface area that lipase enzymes can then work on. This process speeds up their digestion.

Bile is also an alkaline substance. It neutralises any excess stomach acid at the beginning of the small intestine. This process provides the enzymes in the small intestine with their optimum pH.

The lock and key theory

Enzymes are specific for their substrates like keys are specific for their locks. So, protease enzymes will not break down lipids, just as the key to your house will not open your parent's car. For an enzyme to break down a substrate, the substrate must fit into the enzyme, just like a key fits into a lock. So, the shape of the enzyme and substrate must match, just like keys and locks. This model is called the **lock and key theory**.

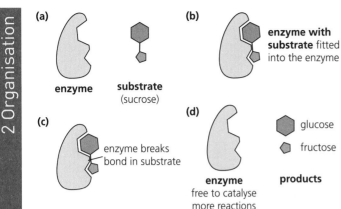

Figure 2.6 **How a digestive enzyme breaks down a substrate. Here the substrate is sucrose and the products are glucose and fructose.**

At optimum pH and temperature, the shapes of the enzyme and substrate fit together perfectly. When we move away from the optimum pH or temperature, the shape of the **active site** changes. This change makes it harder for the enzyme and substrate to fit together and so slows the rate at which the enzyme works. This in turn slows the reaction. If extremes of pH or temperature are reached, the shape of the active site is permanently changed. The enzyme's active site becomes **denatured** and will no longer function.

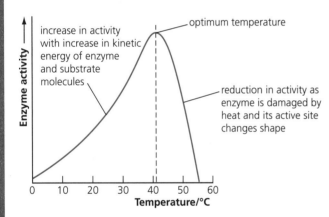

Figure 2.7 **This graph shows the effect of temperature on the activity of an enzyme.**

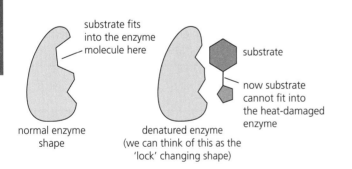

Figure 2.8 **Extremes of temperature and pH denature enzymes by altering the shape of their active site, so the substrate can no longer fit.**

> **Exam tip**
>
> You should be able to use the lock and key theory to explain how enzymes work.

> **Active site**: The region of an enzyme that binds to its substrate.
>
> **Denatured**: A permanent change to an enzyme as a result of extremes of pH and temperature that stop it working.

> **Exam tip**
>
> You should be able to relate the activity of enzymes to their temperature and pH.

> **Exam tip**
>
> You should be able to calculate the rate at which chemical reactions occur.

> **Exam tip**
>
> As well as the lock and key theory, enzyme activity at higher temperatures can be explained by particle theory. At higher temperatures, molecules have more kinetic energy so move faster. This means that they are more likely to collide with substrates.

Required practical 5

pH and rates of reaction

Aim: To investigate the effect of pH on the rate of reaction of amylase enzyme.

Equipment: Boiling tubes, iodine solution, amylase solution, pH buffer solutions, water bath, pipettes, spotting tile.

Method:
1 Place one drop of iodine into each well of a spotting tile.
2 Put $10\,cm^3$ of starch solution in a boiling tube.
3 Place $2\,cm^3$ of amylase solution in another boiling tube.
4 Add $5\,cm^3$ of pH buffer solution to the second boiling tube.
5 Place both boiling tubes into a water bath at 37 °C for 2 minutes.
6 After 2 minutes, add the contents of both boiling tubes together.
7 Every 30 seconds, use a pipette to put one drop of solution into a new well of the spotting tile.
8 Repeat every 30 seconds until the solution turns blue or black.
9 Repeat with different pH buffer solutions.

Results: The longer the time that the iodine test gave positive results, by turning blue or black, the less suitable the pH is to the enzyme. So, the optimum pH is the one at which the solution remained brown for the longest.

> **Vein**: A large blood vessel that returns blood to the heart.
>
> **Artery (plural arteries)**: A large blood vessel that takes blood from the heart.

The heart and blood vessels

REVISED

Your circulatory system is made from your heart and all your **veins**, **arteries** and **capillaries**. Its function is to provide all the millions of billions of cells in your body with the substances they need. All cells need glucose and oxygen for respiration. All cells need carbon dioxide and water (the products of respiration) to be removed. Other cells require different substances such as **hormones** at specific times, such as during puberty. Your circulatory system ensures that all your cells have all the substances they need.

> **Capillaries**: Tiny blood vessels found between arteries and veins that carry blood into tissues and organs.
>
> **Hormone**: A chemical (produced in a gland in mammals) that moves around an organism to change the function of target cells, tissues or organs.

The heart and double circulation

Your heart is an organ made from muscle and nervous tissue. It pumps roughly every second and so your heart beats an amazing 100 000 times a day or over two billion times in your lifetime. Amazingly, many other mammals have heartbeats that are roughly similar and tend to beat around one billion times in their lifetime.

Your heartbeat is controlled by your 'natural pacemaker'. This is a small group of nervous tissue cells in the top right chamber

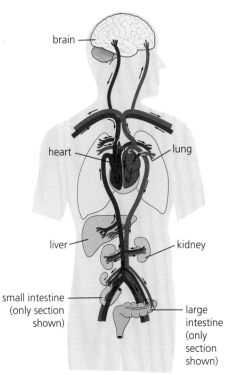

Figure 2.9 The circulatory system, showing the main organs with which the blood exchanges substances. A bright red colour indicates oxygenated blood and a dark red colour indicates deoxygenated blood.

of your heart. They generate an electrical signal which spreads out along your heart's nerve fibres and causes the heart muscle to contract. This action pumps blood from your heart. Your pacemaker controls the rate at which your heart beats.

Blood takes about 1 minute to complete a full circuit of your body. During this time, it will pass through your heart twice. Blood is pumped in a circle from your heart to your lungs, back to your heart and then to the rest of your body, before returning to your heart to start again. This is called double circulation.

Your heart has four chambers: two on each side. The top two chambers are called atria, or your left **atrium** and your right atrium. The bottom chambers are called **ventricles**. Atria and ventricles are separate by valves. Blood collects in your atria when the valves are closed. When your heart beats, the blood in your atria is forced into the ventricles. More valves at the ends of the ventricles stop blood being pumped straight into your blood vessels. When your heart beats again, the blood in your ventricles is forced into your blood vessels to begin another journey around your body.

> **Atrium (plural atria)**: An upper chamber of the heart surrounded by a smaller wall of muscle.
>
> **Ventricle**: A lower chamber of the heart surrounded by a larger wall of muscle.

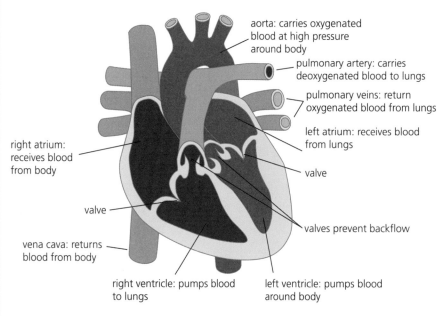

Figure 2.10 A cross-section of the heart, showing its structure.

> **Typical mistake**
>
> Convention dictates that labelled heart diagrams in all books (including this one) have the right-hand side on the left and the other way around.

Blood is pumped from your left atrium into your left ventricle. It is then pumped into an artery called the aorta. The aorta divides into smaller arteries and eventually into tiny capillaries that permeate your tissues and organs to provide oxygen, glucose and other required substances. Your capillaries collect into small veins that join to become larger and eventually form one vein called the vena cava. This vein returns blood to your right atrium. From here it is pumped into your right ventricle and then into an artery called the pulmonary artery. This artery takes blood to your lungs to replenish the oxygen and remove carbon dioxide. Your blood completes its journey by returning to the left atrium in your pulmonary vein.

The muscular lining of your heart is thicker on the left-hand side. This difference is because the left ventricle pumps blood to all tissues of your body including those in your extremities. The right ventricle only pumps blood to the lungs so does not need such a thick muscular lining.

> **Exam tip**
>
> You should be able to describe the structure and function of the human heart and lungs including their adaptations.

The blood vessels

You have between 50 000 and 100 000 miles of blood vessels in your body, which is enough to circle the Earth. There are three types: arteries, veins and capillaries.

Arteries take blood away from the heart. This blood is under high pressure, so the lining of arteries needs to be thick and muscular. It is also elastic so it can stretch when blood is pumped from your heart. You can feel this surge of blood in places such as your wrist where arteries are near the surface. This surge is called your pulse.

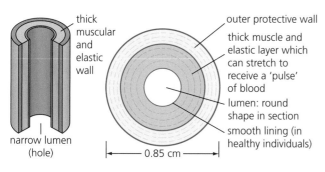

thick muscular and elastic wall

narrow lumen (hole)

outer protective wall

thick muscle and elastic layer which can stretch to receive a 'pulse' of blood

lumen: round shape in section

smooth lining (in healthy individuals)

0.85 cm

Figure 2.11 Note the thickness of the artery wall.

Capillaries carry blood into your tissues and organs to provide all your cells with the substances they need. Capillaries are very small and spread out into your tissues and organs like the roots of a tree. They are very thin. Blood plasma passes from capillaries into the tissues. Here plasma is called tissue fluid and it provides cells with glucose and oxygen for respiration and other substances. Glucose and oxygen move into the cells by diffusion. The products of respiration are carbon dioxide and water. Carbon dioxide and water diffuse into the tissue fluid to be carried away in the capillaries.

Veins carry blood back to the heart under much lower pressure. Blood loses pressure during its journey through the capillaries. So, the linings of your veins do not need to be as muscular as the linings of your arteries. Veins also have one-way valves (not found in arteries) to keep blood flowing back to your heart.

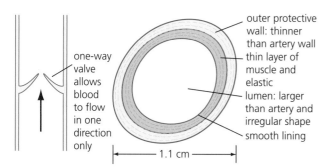

one-way valve allows blood to flow in one direction only

outer protective wall: thinner than artery wall

thin layer of muscle and elastic

lumen: larger than artery and irregular shape

smooth lining

1.1 cm

Figure 2.12 Note the irregular shape and the thinner muscle and elastic layer of the vein.

Typical mistake

The same volume of blood that leaves your heart must return to it. So even though images of veins and arteries show them to have differently shaped structures, the total area of the arteries leaving and the veins returning blood must be the same.

Exam tip

You should be able to explain how the structure of blood vessels relates to their function.

Exam tip

You should be able to calculate the rate of flow of blood.

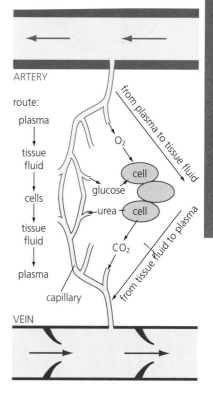

ARTERY

route:
plasma
↓
tissue fluid
↓
cells
↓
tissue fluid
↓
plasma

from plasma to tissue fluid

O_2

cell

glucose

urea — cell

CO_2

from tissue fluid to plasma

capillary

VEIN

Figure 2.13 Exchange between the blood and tissue cells in a capillary network.

Components of blood

You have about 5 litres of blood in your body. Blood looks red because it carries millions of red blood cells around your body. Blood also carries other cells and substances including those absorbed into your blood by your digestive system. These substances include glucose, fatty acids, glycerol and amino acids.

Red blood cells

There are millions of red blood cells in every drop of your blood. These cells carry oxygen from your lungs to your tissues and organs where it is needed for respiration. Red blood cells are a characteristic **biconcave** shape. (They have dips in their middle on both sides.) This shape increases the surface area of the cell, and so increases its ability to absorb oxygen.

Red blood cells contain a compound called **haemoglobin**. In your lungs, this binds with oxygen to form **oxyhaemoglobin**. This process turns the colour of your blood from dark red to bright red. When these red blood cells reach your tissues, they release the oxygen and so oxyhaemoglobin turns back to haemoglobin.

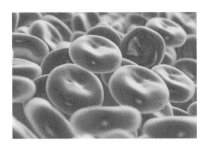

Figure 2.14 The biconcave shape of red blood cells maximises their surface area to volume ratio and so increases their ability to absorb oxygen.

Biconcave: Describes a shape with a dip that curves inwards on both sides.

Haemoglobin: The molecule in red blood cells that can temporarily bind with oxygen to carry it around your body.

Oxyhaemoglobin: A substance formed when haemoglobin in your red blood cells temporarily binds with oxygen.

Phagocyte: A type of white blood cell that engulfs and destroys pathogens.

Lymphocyte: A type of white blood cell that produces antibodies to help clump pathogens together to make them easier to destroy.

Antigen: A protein on the surface of a pathogen that your antibodies can recognise as foreign.

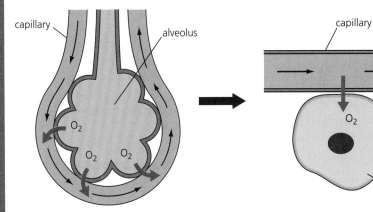

haemoglobin + oxygen → oxyhaemoglobin

oxyhaemoglobin → oxygen + haemoglobin

Figure 2.15 Haemoglobin transports oxygen from the lungs to other organs as oxyhaemoglobin in red blood cells.

White blood cells

White blood cells are part of your immune system to attack pathogens before they can make you ill. You have far fewer white blood cells than red ones. A single drop of your blood will only contain tens of thousands of white blood cells.

There are several types of white blood cells, but two main groups:

Phagocytes engulf pathogens by surrounding them and absorbing them. Break-down enzymes, present inside phagocytes, destroy pathogens. Phagocytes act non-specifically.

Lymphocytes make antibodies that bind to protein **antigens** found on pathogens and, in doing so, clump them together. This process makes it easier for pathogens to be destroyed by phagocytes. Each lymphocyte makes a specific type of antibody that binds to a specific protein antigen on the pathogen. Therefore different lymphocytes are specific to different pathogens.

Platelets

These are not cells, but fragments of them. There are over a 100 000 of them in a drop of blood. They join together after you have cut yourself and form a scab. When you cut yourself, **platelets** release clotting factors. These factors convert a chemical called fibrinogen in your blood into fibrin. This forms a mesh and sticks platelets to it to form a scab.

Plasma

Blood plasma is the straw-coloured fluid that carries red and white blood cells, platelets and all other substances around your body. Just over half of your body's total blood volume is plasma and nearly all your plasma is made of water.

> **Platelets**: Small structures (not cells) in your blood that fuse together to form a scab.

Exam tip

You should be able to describe the functions of all blood components. You should also be able to recognise the different types of blood cell in a diagram and explain how their structure is related to their function.

Coronary heart disease: a non-communicable disease

REVISED

Communicable diseases like the common cold can be transmitted between people. Coronary heart disease is a non-communicable disease. This means it develops, rather than being caught.

Your heart is like all other organs in your body. Its cells need glucose and oxygen for respiration. These are delivered to your heart in its **coronary arteries**.

Unhealthy lifestyles include not doing enough exercise, smoking, drinking in excessive amounts and a poor diet. This can result in the build-up of fat inside coronary arteries and also a reduction in the flexibility of their linings. This process is called **atherosclerosis**. Fat build-up and atherosclerosis slow or stop blood from reaching the heart. This can result in heart attacks.

Coronary heart disease is now one of the major causes of death in the world. A common previous treatment for this was a **heart bypass** operation. This treatment involved moving a short section of artery from another part of a person's body and using it to short circuit the blockage. This is a major form of surgery and has inevitable risks. More recent treatments are less damaging. **Stents** are small medical meshes that can be expanded to hold open blocked arteries. Statins are drugs prescribed by doctors to reduce **cholesterol**, which in turn reduces coronary heart disease.

> **Communicable**: A disease that can be transmitted from one organism to another.
>
> **Coronary arteries**: Arteries that supply the heart muscle with oxygenated blood.
>
> **Atherosclerosis**: A medical condition resulting from an unhealthy lifestyle that reduces the flexibility of arteries.
>
> **Heart bypass**: A medical procedure in which a section of less important artery is moved to allow blood to flow around a blockage in a more important one.
>
> **Stent**: A small medical device made from mesh that keeps arteries open.
>
> **Cholesterol**: An important biological molecule for cell membranes, which leads to atherosclerosis if found in high levels in the blood.

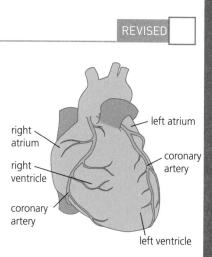

Figure 2.16 The coronary arteries supply glucose and oxygen to the heart muscle tissue.

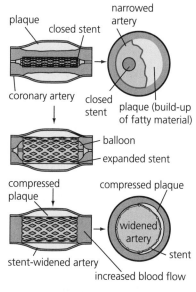

Figure 2.17 This stent allows blood to flow freely again.

Faulty valves

Your heart has valves between the atria and ventricles and also between the ventricles and arteries. Some heart valves do not function properly. If they do not function properly, less blood will be pumped around a person's body. Valves can be replaced in a major heart operation. They can be from organ donors or artificial, mechanical values.

Pacemakers

Some people have difficulties controlling their heartbeat. The beat needs to be coordinated properly to ensure the maximum volume of blood is pumped at each heartbeat. An artificial pacemaker can be fitted to people in this situation. The pacemaker takes over the generation of the electrical signal needed to coordinate every heartbeat.

Transplants

If all other treatments fail, some patients need a heart transplant. These treatments are serious medical procedures. All transplants, including this type, need a match between the donor and that of the patient. Some patients can sadly be on waiting lists for a heart transplant for many years.

> **Exam tip**
>
> You should be able to evaluate the advantages and disadvantages of the various cardiovascular treatments listed above.

Health issues

REVISED

Your health is now defined as your mental and physical wellbeing. It can be improved by:
- seeking medical help when you are physically or mentally unwell
- eating a balanced diet
- exercising regularly
- reducing stress levels.

Well-balanced diet

A well-balanced diet has the correct amounts of all food groups. This can be shown in a food pyramid. This model shows that a large part of your diet should be made from carbohydrates like bread, potatoes and rice. Another large part of your diet should be made from fruits and vegetables. These contain natural healthy sugars, vitamins and fibre. Fats are found in dairy products including milk and cheese. These are also high in protein. They do however, have high cholesterol levels and so should be eaten in smaller quantities.

Figure 2.18 A balanced food diet represented as a food pyramid.

Regular exercise

Exercise strengthens your muscles including your heart. It also improves the efficiency of your circulatory system. Many people also believe that exercise helps your mental, as well as physical, health. The National Health Service (NHS) recommends that all young people do an hour of exercise every day.

Physical and mental ill health

Doctors now believe that there are links between diseases and mental and physical health. Patients with one medical condition can be more likely to develop another:
- If a person's immune system is not functioning properly, they are more likely to catch communicable diseases.
- A small number of viral infections leads to cancer.
- Serious physical health problems can lead to mental ill health.

Stress is the feeling of being under too much mental or emotional pressure. Anxiety is the feeling of unease, which is often worry or fear. Depression affects different people in many ways including physical ones like losing appetite. It is crucial that anyone who feels stressed, anxious or depressed speaks to their doctor as soon as they can.

Cancer

Sometimes the process of cell division goes wrong and forms cancerous cells. These multiply in an uncontrolled way and can form a lump or tumour. **Malignant** tumours are cancerous. They can spread to other parts of the body and form secondary tumours in a process called **metastasis**. It is essential that malignant tumours are removed in an operation or that their cells are killed before they spread. **Benign** tumours are less serious. They often are found within a membrane and so are less likely to spread.

Sadly, it is likely that one in every two people will suffer from cancer at some point in their life. Symptoms include a lump, weight loss and unexplained bleeding. There are over 200 types of cancer, many of which have different symptoms. It is crucial that anyone with concerns speaks to their doctor as soon as they can.

Screening

Screening is the medical process of looking for cancers. Doctors feel for lumps, take blood and urine tests and X-rays during screening. Women often have two additional screening tests. **Mammograms** are X-rays that are automatically offered by the NHS to women between 50 and 70 years of age to screen for breast cancer. **Smear tests** look for cervical cancer. Cells are collected from the **cervix** and are viewed under a microscope to check for cancerous cells. The NHS automatically offers smear tests to women between 25 and 49 years of age every 2 years, and less frequently beyond the age of 50.

Causes of cancer

More than 20% of cancers are caused by smoking. Excessive drinking also causes many cancers. Hepatitis B and C virus infections and human papilloma virus (HPV) can also cause cancer. **Ionising radiation** from the Sun's ultraviolet (UV) rays and chemical pollutants from factories can cause cancer. The older a person gets the more likely they are to suffer from cancer.

Treating and preventing cancer

Chemotherapy and radiotherapy are the two most common forms of cancer treatment. They are often used together. Chemotherapy uses chemicals to kill cancerous cells. Radiotherapy uses X-rays to do the same. With both treatments, it is often difficult for doctors to only kill cancerous cells. Side effects of both treatments include hair loss, being tired and being sick.

The development of cancers is prevented by leading a healthy lifestyle. This includes not smoking, drinking in low amounts or not at all, exercise and a balanced diet.

The effect of lifestyle on some non-communicable diseases

Risk factors are aspects of your lifestyle or substances taken into your body that increase the risk of a disease occurring. Some risk factors are known to cause disease. We call this

> **Exam tip**
>
> You should be able to describe cancer as the result of changes in cells that lead to uncontrolled growth and division.

> **Malignant**: A cancerous tumour that can spread to other parts of the body.
>
> **Metastasis**: The development of tumours following the spread of malignant cancer.
>
> **Benign**: A non-cancerous tumour that does not spread.
>
> **Mammogram**: A medical procedure using X-rays to check for breast cancer.
>
> **Smear test**: A medical procedure to check for cancer of the cervix.
>
> **Cervix**: The narrowing between the vagina and the uterus in the female reproductive system.
>
> **Ionising radiation**: UV rays, X-rays and gamma rays that can cause **mutations** to DNA.
>
> **Mutation**: A permanent change to DNA, which may be advantageous, disadvantageous or have no effect.

causation. Others are only associated with diseases. Here no cause had been proved. This is called **correlation**. Table 2.4 shows risk factors, diseases and their effects.

Table 2.4 Risk factors, diseases and their effects.

Risk factor	Disease	Effects
Obesity and lack of exercise	Type 2 diabetes	Body cells do not respond to the hormone insulin, which helps control the glucose level in the blood.
Alcohol	Liver function	Long-term alcohol use causes liver cirrhosis. The cells in the liver stop working and are replaced by scar tissue. This stops the liver from removing toxins, storing glucose as glycogen and making bile.
Alcohol	Brain function	Excessive use of alcohol can also alter the chemicals in the brain (neurotransmitters), which pass messages between nerve cells. This can cause anxiety and depression, and reduced brain function.
Smoking	Lung disease and cancer	Smoking can cause cancer in many parts of the body, including the lungs, mouth, nose, throat, liver and blood. It also increases the chances of having asthma, bronchitis and emphysema.
Smoking and alcohol	Underdevelopment of unborn babies	Alcohol and chemicals from cigarettes in the mother's blood pass through the placenta to her baby. Without a fully developed liver the baby cannot detoxify these as well as the mother can. This can lead to miscarriage, premature birth, low birth weight and reduced brain function.
Carcinogens and ionising radiation	Cancer	Chemicals and radiation that cause cancer are called **carcinogens**. Tar in cigarettes, asbestos, ultraviolet from sunlight and X-rays are examples.

Exam tips

- You should be able to describe the relationship between health and disease and the interactions between diseases shown in Table 2.4.
- You should be able to describe data about diseases in frequency tables and diagrams, bar charts and histograms, and use a scatter diagram to identify correlations.
- You should be able to discuss the human and financial cost of non-communicable diseases.
- You should be able to explain the effect of lifestyle factors on the levels on non-communicable disease.

Risk factor: Any aspect of your lifestyle or substance in your body that increases the risk of a disease developing.

Causation: The act of causing an outcome.

Correlation: When an action and outcome are linked, but the action does not cause the outcome.

Now test yourself

TESTED ☐

1 Describe the process of peristalsis.
2 What are the three enzymes of the human digestive system?
3 Describe the food test for starch?
4 Describe what occurs when an enzyme is denatured.
5 Why is the human circulatory system often described as double circulation?
6 How are veins adapted for their function?
7 What are the four components of blood?
8 What is the function of lymphocytes?
9 Why are stents used more commonly for coronary heart disease than bypass operations?
10 How can you improve your health?

Answers on p. 113

Revision activity

Draw out Table 2.4 with only the headings along the top and the first column on the left. Try to fill in the rest of the table from memory to help you to revise.

Plant tissues, organs and systems

Plant tissues

Just as with you and other animals, multicellular plants have several levels of organisation. Cells are the basic building blocks of all life. Tissues are groups of cells with a similar structure and function.

Epidermal tissue

The outside layer of a plant is called its **epidermis**. This layer of cells has many functions including protection against the loss of water, the exchange of gases between the leaves and the air and the uptake of water in the roots. The epidermis is transparent to allow light to pass through it for **photosynthesis**.

Palisade mesophyll

Below the epidermis in leaves is the **palisade mesophyll** layer. This layer contains cells with very high numbers of chloroplasts to maximise the amount of glucose produced in photosynthesis.

Spongy mesophyll

Below the palisade mesophyll layer in leaves is the **spongy mesophyll** layer. These cells have fewer chloroplasts because they are further from sunlight. The cells of this layer are less regularly shaped and have gaps in between them. These gaps allow gases to diffuse from within the leaf into the air and the reverse.

Small pores called **stomata**, which are mainly found on the bottom of leaves, open and close to allow gases, including water vapour, to diffuse into and out from leaves at different rates. Stomata are surrounded by guard cells that swell to open the stomata and shrink to close it.

Xylem and phloem

Xylem tubes carry water and mineral ions from the roots to the leaves in a process called **transpiration**. Xylem tissue is made from hollow tubes strengthened by lignin to transport water to the leaves for photosynthesis.

The sugars made in photosynthesis dissolve in water and are transported around the plant in phloem tubes for respiration or storage in a process called translocation. Phloem tissues are made from tubes of elongated cells. Cell sap can move from one cell to the next through pores.

> **Epidermis**: The outermost layer of cells of an organism.
>
> **Photosynthesis**: A chemical reaction that occurs in the chloroplasts of plants and algae to store energy in glucose.

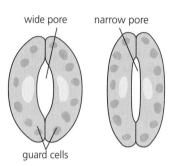

Figure 2.19 **Pores called stomata become smaller if a plant needs to reduce water loss by transpiration.**

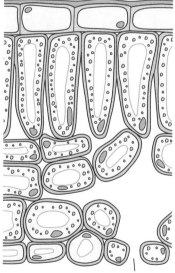

Figure 2.20 **Part of a leaf in cross-section showing the epidermis (top and bottom layers), palisade mesophyll cells (long, thin upright cells below the upper epidermis) and the spongy mesophyll cells (more circular cells found towards the bottom of the leaf).**

> **Palisade mesophyll**: Cells found towards the top of leaves with lots of chloroplasts for photosynthesis.
>
> **Spongy mesophyll**: Cells found towards the bottom of leaves with spaces in between them to allow gases to diffuse.
>
> **Stomata (singular stoma)**: Tiny holes in leaves bordered by guard cells that allow gases to diffuse in and out.
>
> **Xylem**: Dead plant cells joined together into long tubes through which water flows during transpiration.
>
> **Transpiration**: The gradual release of water vapour from leaves to continue the 'pull' of water up to them from the soil.

Xylem and phloem tubes are often found together in structures called vascular bundles.

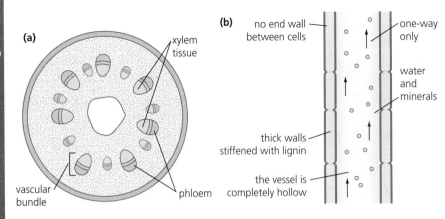

Figure 2.21 (a) Transverse section of a stem, (b) longitudinal section of xylem vessel from a vascular bundle (the arrows show the direction of water flow).

Meristem

The tips of roots and shoots have a region of rapidly growing cells called the meristem. Here mitosis happens.

Plant organs

REVISED ☐

Just like in you and other animals, plant organs are groups of tissues that perform a specific function.

Root

Roots grow downwards towards water. They are usually white because they possess no chloroplasts. Cells that spend their lives underground do not need to photosynthesise. Roots anchor plants in the soil, between rocks or on trees. Some roots store excess glucose made in photosynthesis as starch. Root hair cells have an extension protruding into the soil. This massively increases their surface area, meaning they can absorb more water by osmosis and minerals by active transport.

Shoot

A shoot is the newly formed stem, its leaves and any buds. Just like the tips of roots, shoots also possess a meristem. Here cells divide by mitosis so the shoot continues to grow towards the light. (Root meristems allow roots to continue to grow downwards towards water.)

Leaf

Leaves are plant organs. They are the major site of photosynthesis. Evergreen plants retain their leaves throughout the year. **Deciduous** plants drop their leaves in autumn when light levels are too low to keep them. They regrow in the spring when light levels increase again.

Deciduous: Describes broadleaved trees that drop their leaves in winter.

Plant organ systems

Just like in you and other animals, plant organ systems are groups of organs with similar functions. Plants are made from their organ systems.

Transportation organ system

The transportation organ plant system is made from the roots, shoots and leaves of plants. (These are all organs.) This system acts like your circulatory system and moves all the substances the plant needs around their structures.

Translocation

Phloem tubes carry sugars and minerals dissolved in water throughout a plant. Sugars are made during photosynthesis and are broken down to release energy during respiration.

Transpiration

Plants do not have a heart so they cannot pump water to their leaves for photosynthesis. Instead we say it is 'pulled' upwards in a transpiration stream. Water is purposefully allowed to evaporate from the stomata. Guard cells control how quickly or slowly this happens. As the water evaporates, more is pulled up from the roots to replace it. This process therefore occurs continuously throughout the life of all plants.

Transporting water by transpiration allows plant cells to remain **turgid**. Transpiration is vital for photosynthesis and also enables essential minerals to be carried from the soil.

Transpiration rates increase when:
- There is more wind.
- The air is drier (less **humid**).
- The temperature is higher.
- The light intensity is higher (on sunny days).

> **Exam tip**
>
> You should be able to calculate the rate of transpiration. You should then be able to explain the effect of changing temperature, humidity, air movement and light intensity on the rate of transpiration.

> **Exam tip**
>
> You should be able to describe the processes of transpiration and translocation, including the structure of stomata.

> **Turgid**: Describes swollen cells.
>
> **Humid**: Describes an atmosphere with high levels of water vapour.

Now test yourself

TESTED

11 What are plant tissues?
12 What cell component is present in high numbers in the palisade mesophyll cells and why?
13 Why are there spaces between cells in the spongy mesophyll layer?
14 When a plant is respiring, what gases move out from the leaves?
15 Where are guard cells found?
16 Describe the process of transpiration.
17 What term describes the growth of roots downwards?
18 Describe the process of translocation.
19 What term do we use to describe both xylem and phloem tubes close together in a plant?
20 What is different about xylem and phloem tubes?

Answers on p. 113

Summary

- Cells are the basic building blocks of all living organisms. Tissues are groups of cells with a similar structure and function. Organs are groups of tissues performing specific functions. Organs are arranged into organ systems which make up organisms.
- The digestive system is an organ system that breaks down large lumps of insoluble food, into smaller, soluble molecules, which are absorbed into the blood.
- Enzymes are biological catalysts that speed up reactions. An enzyme's substrate fits into its active site. The lock and key theory is a model to explain that enzymes are specific for their substrates.
- Enzymes have an optimum pH and temperature at which their activity is greatest. At extremes of pH or temperature, the shape of the active site changes and the enzyme becomes denatured. The substrate will no longer fit and so the enzyme will no longer work.
- Carbohydrase enzymes break down carbohydrates into sugars. Lipase enzymes break down fats and oils (lipids) into fatty acids and glycerol. Protease enzymes break down proteins into amino acids. The pancreas and small intestine make all three enzymes. The mouth makes carbohydrase enzymes. The stomach makes protease enzymes. Bile is not an enzyme, but it emulsifies fats. It is made in the liver.
- The human heart pumps blood around the body in a double circulation system. The heart is made from four chambers. The atria (upper chambers) pump blood to the ventricles, which then pump blood either to the lungs (right ventricle) or to the rest of the body (left ventricle).
- The lungs provide the blood with oxygen and remove carbon dioxide.
- Arteries, like the aorta and the pulmonary artery, take blood away from the heart. Arteries divide into tiny capillaries, which permeate tissues. Veins, like the vena cava and the pulmonary vein, return blood to the heart.
- Blood is a tissue. It contains red blood cells that carry oxygen, white blood cells that are part of the immune system and platelets that are cell fragments that form scabs. These are all carried in blood plasma.
- Coronary heart disease occurs when layers of fat block the inside of the arteries supplying the heart. The fat reduces the supply of oxygen and glucose needed for respiration. Bypass operations, the insertion of stents and statin drugs are all used to treat coronary heart disease. Heart and lung transplants can replace faulty organs.
- Health is the state of physical and mental wellbeing. Diseases are a major cause of ill health. Other factors like diet, stress and life situations can cause poor physical and mental health. Different types of diseases can interact.
- Risk factors are linked to an increased rate of a disease. Risk factors can be part of a person's lifestyle or substances in the body or environment. Causation occurs when risk factors have a proven link to disease. An example of this is smoking and lung disease and cancer.
- Cancer is the result of changes in cells that lead to rapid cell growth. Benign tumours are growths of abnormal cells that do not spread. Malignant tumours are cancers that do spread.
- Plant tissues include epidermal tissues, palisade mesophyll, spongy mesophyll, xylem and phloem and meristem tissue at the tips of roots and shoots. The leaf is a plant organ.
- Transpiration is the movement of water from the soil through root hair cells, up xylem tubes to the leaves. Plants release water vapour through stomata in their leaves to ensure transpiration occurs continuously. Transpiration rate is affected by temperature, humidity, air movement and light intensity.
- The roots, stem and leaves form the plant transportation organ system.
- Translocation is the movement of dissolved sugars made during photosynthesis from the leaves to the rest of the plant in phloem cells.

Exam practice

1 What type of enzyme is present in saliva? [1]
   A Carbohydrase
   B Lipase
   C Protease
   D Bile
2 What name is given to the process of moving sugary water from the leaves of a plant? [1]
   A Transpiration
   B Photosynthesis
   C Osmosis
   D Translocation
3 Define the term artery. [1]
4 Describe the pathway of blood around the body beginning with the left atrium. (Specific names of blood vessels are not required.) [4]
5 Describe the hierarchy of organisation in plants. [4]
6 Describe the shape of this graph and explain its significance. [4]

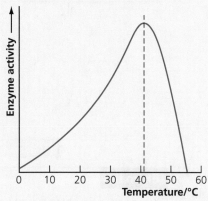

7 Describe how you investigate the effect of pH on enzyme activity of carbohydrase enzyme. [6]
8 Explain how plant roots are adapted for osmosis and active transport and give an example of a substance absorbed by each process. [6]
9 Explain how the lock and key theory models enzyme action including denaturing. [6]

Answers and quick quizzes online

ONLINE

3 Infection and response

Infection and response

Communicable (infectious) diseases

REVISED

Pathogens are micro-organisms that pass disease from one organism to another. Diseases which can be passed by pathogens are called communicable (or infectious). There are four types of pathogen:
- viruses, such as measles
- bacteria, such as *Salmonella*
- fungi, such as rose black spot in plants
- protists, such as **malaria**.

The lifecycle of all pathogens is similar:
1 They infect a host.
2 They reproduce (or replicate if a virus).
3 They spread from their host.
4 They infect another host and repeat.

> **Malaria**: A communicable disease, caused by a protist transmitted in mosquitos, which attacks red blood cells.

Spread of pathogens

REVISED

Pathogens spread disease in the following ways:
1 Airborne: when people sneeze, the virus that causes the common cold can be spread through the air in tiny droplets.
2 Direct contact (sexual or non-sexual): the *Chlamydia* bacterium is the cause of a sexually transmitted disease (STD) that passes from one person to another during sex.
3 Dirty water: cholera is caused by a bacterium that is spread in dirty water.
4 Contaminated food: food poisoning is often caused by the *Escherichia coli* bacterium in undercooked or reheated food.
5 **Vectors**: some farmers believe that the tuberculosis bacterium is passed from badgers to their cows. Any animal, or micro-organism that does this is called a vector.

The spread of disease can be reduced or prevented by:
- high levels of personal hygiene including hand washing
- covering your mouth and nose when you cough or sneeze
- cleaning and disinfecting surfaces and objects with antiseptics
- **vaccination** and taking medicines when prescribed
- avoiding close contact with people who are sick.

> **Exam tip**
>
> You should be able to explain how diseases caused by viruses, bacteria, protists and fungi are spread in animals and plants. You should also be able to explain how this spread can be reduced.

> **Vector**: An animal that transmits a communicable disease without being infected itself.
>
> **Vaccine**: A medicine containing an antigen from a pathogen that triggers a low level immune response so that subsequent infection is dealt with more effectively by the body's own immune system.

Viral diseases

REVISED

Viruses are not considered to be alive because they do not fulfil all the seven life processes. For example, they replicate not reproduce and they do not respire. They are therefore classified as strains and not **species**. They are some of the smallest pathogens, made from short lengths of DNA or RNA surrounded by a protein coat. They infect individual cells and use the host cell to replicate. This causes the cell to burst, allowing new viruses to infect surrounding cells.

> **Species**: The smallest group of classifying organisms, all of which are able to interbreed to produce fertile offspring.

Measles

Measles is caused by a virus and can be fatal if complications occur. Most young children are vaccinated against measles. Its symptoms include a fever and red skin rash. It is spread through the air in droplets from coughs and sneezes.

HIV/AIDS

Human immunodeficiency virus (HIV) is a virus that is spread by sexual contact during exchange of bodily fluids. This can also occur when blood is swapped in shared needles used by drug users. Initially an infected person will feel flu-like symptoms. Unless an infected person is given antiviral drugs, the virus will attack their body's immune system. HIV disease is called AIDS (acquired immune deficiency syndrome) when the person's immune system can no longer defend them.

Tobacco mosaic virus

The tobacco mosaic virus infects many plants including tomatoes, not just tobacco. You can learn about this in the section on plant disease.

Bacterial diseases

REVISED

Bacteria are alive. They are prokaryotes, so do not have a nucleus. Not all bacteria cause disease. Many, including those in your digestive system, are useful to us. Bacteria live on or in living organisms and are often found in places such as mouths, noses and throats. They do not need the body's individual cells to live, as viruses do. Pathogenic bacteria may produce poisons (toxins) that damage tissues and make us ill.

Salmonella

The *Salmonella* bacterium causes food poisoning. It is often spread in food prepared in unhygienic conditions, or that is undercooked or reheated. Symptoms of infection include fever, abdominal (tummy) cramps, vomiting and diarrhoea. All poultry in the UK are vaccinated against *Salmonella*.

Gonorrhoea

Gonorrhoea is a sexually transmitted disease (STD) spread by sexual contact. It causes a thick yellow or green discharge from the vagina or penis. Urinating is painful. Initially the bacterium that causes the infection was easily treated by penicillin (the first antibiotic). Latterly it has evolved to be resistant to this drug. Treatment is now by more recently discovered antibiotics. Its spread can be reduced by barrier contraception such as a condom.

Fungal diseases

REVISED

Fungi are eukaryotes like animals and plants. Their cells have a nucleus. They have cell walls like plants, but these are made from **chitin** and not cellulose. They can be singled-celled like yeast or multi-celled like mushrooms. Not all fungi cause disease. Some like yeast are very important economically in making bread and beer.

> **Chitin**: A polymer made from sugars that forms the cell walls of fungi and the exoskeleton of insects.

Rose black spot

Black spot is a fungal pathogen that infects roses. You can learn about this in the section on plant disease.

Protist diseases

Protists are also eukaryotes. They are single- or multi-celled but they do not have tissues, organs nor organ systems like animals, plants and fungi.

Malaria

One of the most common diseases caused by a protist is malaria. This protist is spread in blood sucked by mosquitos. These are therefore the vector for the spread of the disease. The protist causes recurring fevers that can be fatal if not treated. Reduction in the spread of malaria occurs by preventing mosquitos from breeding and by using mosquito nets and sprays to avoid being bitten.

Human defence systems

The first line of defence

The first line of defence stops pathogens from entering your body. These defences are non-specific. They are described in Table 3.1.

Table 3.1 **Your first line of defence against infection by pathogens.**

Type	Description
Skin	Your skin is an organ. It has many functions including insulation and sensitivity but also provides a barrier that almost completely covers you to prevent attack by pathogens. When your skin is broken with a cut or graze, your body works very hard to form a scab to prevent pathogens entering.
	Skin does not cover your mouth or eyes. Here your body produces enzymes called **lysozymes** that attack bacteria by breaking down their cell walls.
Nose	Your nose has hairs and produces mucus to trap pathogens you might have breathed in.
Trachea and bronchi	The cells that line your airways (trachea and bronchi) possess tiny hair-like projections called **cilia**. In between these ciliated cells are goblet cells which produce mucus. This traps any pathogens that have bypassed the hairs and mucus in your nose. The hairs of the ciliated cells beat in a rhythmical motion to waft the mucus and its trapped pathogens up to your throat. When you clear your throat, you swallow this mucus into your stomach where any pathogens are killed.
Stomach	Your stomach contains hydrochloric acid. This does not break down food directly but is strong enough to kill many bacterial pathogens that enter through your mouth or nose.

Lysozymes: Antibacterial enzymes found in your tears to prevent eye infections.

Cilia: Tiny hair-like projections from ciliated cells that waft mucus out of the gas exchange system.

Revision activity

Draw out this table with only the headings along the top and the first column on the left. Try to fill in the rest of the table from memory to help you to revise.

The second line of defence

If a pathogen passes your first line of defence, it is attacked by your second line of defence. This is again non-specific, so all pathogens are attacked in the same way. White blood cells called phagocytes attack all pathogens that have evaded your first line of defence. Their cell membrane flows around the pathogens engulfing them in a vacuole. Enzymes within the vacuole then attack the pathogen cell walls and membranes. This process is called phagocytosis.

Exam practice answers and quick quizzes at **www.hoddereducation.co.uk/myrevisionnotes**

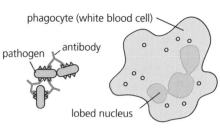
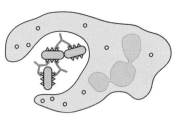

(a) Antibodies cause pathogens to clump together

(b) Phagocyte flows around pathogens to engulf them in a vacuole

(c) Enzymes added to vacuole to break down pathogen cell walls and membranes

Figure 3.1 **A white blood cell (phagocyte) engulfing pathogens.**

The third line of defence

The third line of defence attacks pathogens in a more specific way. Another type of white blood cells called a lymphocyte produces **antibodies** that specifically match proteins called antigens found on the outside of pathogens. Every pathogen has different antigens and so every time you are infected by a pathogen your lymphocytes produce different antibodies. Your lymphocytes 'remember' each pathogen and can produce more of the specific antibodies faster if you are exposed to the same pathogen again. This means you do not catch the same common cold each winter. There are in fact several hundred different common colds. Antibodies cause pathogens to stick together making it easier for phagocytes to engulf and destroy them.

Your lymphocytes also produce **antitoxins**. These are a special type of antibody that can neutralise the toxins produced by some pathogens that make you feel ill.

> **Exam tip**
>
> You should be able to describe the non-specific defence systems and the role of the immune system in the defence against disease.

> **Antibodies**: Proteins produced by lymphocytes that recognise the antigen of pathogens and help to clump them together.
>
> **Antitoxin**: A protein produced by your body to neutralise harmful toxins produced by pathogens.

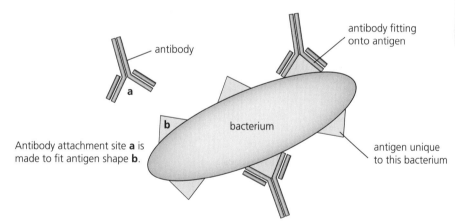

antibody fitting onto antigen

antibody

a

b

bacterium

Antibody attachment site **a** is made to fit antigen shape **b**.

antigen unique to this bacterium

Figure 3.2 **How an antibody fits onto the antigen of a pathogen.**

Vaccination

Vaccinations protect us from disease. If a large proportion of a population is vaccinated it is much less likely that a disease will spread. This is called herd immunity. A vaccination is a small quantity of dead or inactive form of a pathogen. This is introduced into the body, often by injection. This stimulates the lymphocytes (white blood cells) to produce antibodies.

Shortly after a vaccination you may feel a little sick. This is your body fighting the disease and is called the initial exposure. However, if you were to encounter a more severe case of the pathogen in later life (a secondary exposure), then your lymphocytes 'remember' the infection and produce more antibodies faster. In doing so, you are much less likely to fall ill.

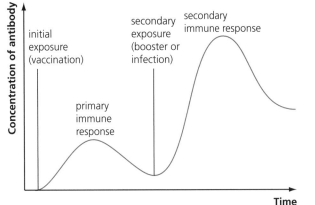

Figure 3.3 **This graph shows the rate at which antibodies are produced after the first and second exposures to a pathogen.**

Antibiotics and painkillers

REVISED

Antibiotics

Antibiotics, such as penicillin, are medicines that cure bacterial disease by killing bacteria within an infected person's body. They do not kill viral pathogens, so you will never be prescribed antibiotics for a common cold (caused by a virus). It is difficult to develop drugs that kill viruses within cells without hurting the cells themselves.

Since the discovery of the first antibiotic by Sir Alexander Fleming (1881–1955) around a hundred years ago, millions of lives have been saved by their use. However, some bacteria are rapidly evolving resistance to our antibiotics. In fact, they are doing this much faster than we can find new ones. Doctors are worried that soon many of our antibiotics will not work anymore and that deaths will increase again.

Painkillers

Painkillers are drugs that reduce or stop pain. Some painkillers were discovered in plants. Aspirin naturally occurs in the bark of willow trees. Others like paracetamol are artificial and have been designed by chemists. Painkillers are used to reduce the symptoms of disease such as pain, swelling and fever. They do not kill the pathogens themselves.

Discovery and development of drugs

REVISED

Drugs are chemicals that have a biological effect on the organism that takes it. Drugs can be natural or man-made. They can be helpful like medicines or harmful like many illegal drugs. Some are more easily obtained like alcohol and tobacco. The use of others is controlled by doctors and can only be taken with a medical prescription.

Traditionally drugs were extracted from plants and micro-organisms. The heart drug digitalis originated from foxglove plants. The painkiller aspirin originated in the bark of willow trees. The antibiotic penicillin was discovered by mistake from the *Penicillium* mould. Most new drugs are developed by chemists in the pharmaceutical industry, although the starting point may be a natural chemical compound.

Modern drug development

Drug development is a long and expensive process that takes many years and costs hundreds of millions of pounds. Only around 0.1% of drugs pass this testing process. The process is shown in Table 3.2.

Table 3.2 The stages of modern drug development

Stage	Description
One	This uses **computer modelling** to look at the structure of the drug and the interactions it might have on the human body.
Two	Stage two involves laboratory tests. These can be on cells grown in the laboratory or on animals. The results from these are used to predict how the drug will affect humans.
Three	The final stage involves human trials. In the first part of this stage, the drug is given to a small number of healthy volunteers. This is to determine the correct dosage. The second round of tests here are given to sick patients to see how effective it is (its **efficacy**). The final stage involves tests on much larger numbers of volunteers to check dosage and efficacy.
	In some of the tests, some of the volunteers are given **placebo** doses. These look the same as the active drugs. They are used to eliminate the placebo effect, which is where people can feel better because they think they have taken the drug even if they actually have not. Some of the tests are called 'double blind'. In these, neither the doctors or the volunteers know if they have been give the drug or the placebo. This eliminates any bias.

Now test yourself

TESTED ☐

1 What are pathogens?
2 What are vectors? Give an example of a vector in your answer.
3 State an example of a communicable disease caused by a virus, bacterium and fungus.
4 What do all examples of the first line of defence stop?
5 What are lysozymes?
6 How are phagocytes and lymphocytes different?
7 What do lymphocytes produce to neutralise harmful toxins produced by pathogens?
8 What is herd immunity and why is it important?
9 What type of pathogen are antibiotics useless to treat?
10 Describe a double-blind trial.

Answers on p. 114

Revision activity

Draw out this table with only the headings along the top and the first column on the left. Try to fill in the rest of the table from memory to help you to revise.

Computer modelling: Using computer software to theoretically examine or test.

Efficacy: The effectiveness of a drug.

Placebo: A medicine that has only psychological effects.

Double-blind trial: A medical experiment in which the patients and doctors do not know who has been given the drug and who has been given the placebo.

Monoclonal antibodies

Producing monoclonal antibodies

REVISED ☐

Antibodies are proteins made by your white blood cells (lymphocytes). They are made to bind to specific pathogens. Antibodies attach themselves to proteins called antigens that are found on the surface of pathogens, and in fact on many other substances. The shape of the antibody fits perfectly with the shape of the antigen.

We have recently developed the ability to make antibodies fit antigens found on other substances not just pathogens. We call these **monoclonal antibodies** and they are potentially extremely useful in medicine. Monoclonal means many copies (clones) of one (mono).

Procedure

1 A mouse is injected with the substances that antibodies are to be made to fit. This must contain antigens.
2 Over the next few days the mouse will naturally make antibodies to attack this substance (as part of its immune response).
3 Small numbers of spleen cells that make antibodies are removed from the mouse in an operation.
4 These spleen cells are fused with cancerous **myeloma** cells. These divide by mitosis in an uncontrolled manner. This process forms **hybridoma** cells. These cells will live indefinitely and continually produce monoclonal antibodies as they do.
5 These monoclonal antibodies will now bind to the original substance injected into the mouse. They are collected by centrifugation, filtration and chromatography.

> **Monoclonal antibodies**: Identical antibodies produced from the fusion of mouse **spleen** cells and myeloma cells.
>
> **Spleen**: An organ of the immune system, found in the abdomen of most vertebrates, that regulates blood cells.
>
> **Myeloma**: A malignant tumour of white blood cells in bone marrow.
>
> **Hybridoma**: A monoclonal antibody-producing cell formed from the fusion of a mouse spleen cell and myeloma cell.

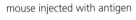

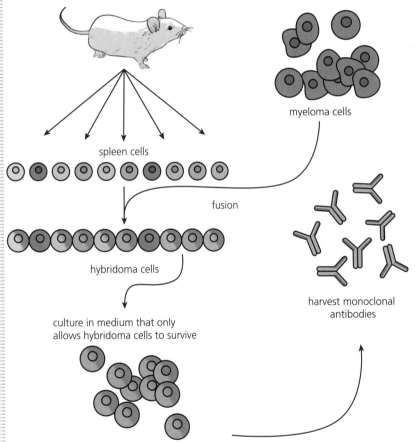

> **Revision activity**
>
> Write out the procedure for making monoclonal antibodies as a flow diagram to help you remember.

Figure 3.4 **The process by which monoclonal antibodies are produced.**

Using monoclonal antibodies

REVISED

Procedure

Monoclonal antibodies are used in pregnancy test kits. They are placed on the ends of a short stick onto which a woman urinates. If she is pregnant she will have a small amount of a specific hormone called human chorionic gonadotropin (hCG) in her urine. This binds to the monoclonal antibodies on the end of the stick and indicates that she is pregnant. If she is not pregnant she will not have hCG hormone in her urine and so it will not bind to the monoclonal antibodies on the stick.

Other uses

Monoclonal antibodies have been designed to bind to and therefore diagnose infections including hepatitis, HIV/AIDS, herpes and chlamydia. We can design monoclonal antibodies to contain a fluorescent dye. Under UV light they will then glow and so identify an area of the body that contains the corresponding antigen.

In fact, monoclonal antibodies could be made to bind (and so diagnose the presence of) almost any substance in the body. We call tests like this **diagnostic**. They diagnose whether there is a medical issue, but do not treat it.

Monoclonal antibodies can also treat illnesses. They have been designed to help your immune system attack cancerous cells and carry toxic drugs or radiation to them to kill them. They can also stop these cells dividing. We can design monoclonal antibodies to only bind to cancerous cells, meaning that the surrounding healthy cells will not be killed.

Shortly after monoclonal antibodies were first designed many people thought they were a 'magic bullet' capable of treating many illnesses. Unfortunately, there have been more side effects in their use than was first anticipated. These have been identified in the rigorous process of drug testing that we now have.

> **Diagnostic test:** A medical procedure that tells a patient whether they have an infection or condition.

> **Exam tip**
>
> You should be able to describe the production and uses of monoclonal antibodies. You are not expected to recall any specific tests or treatments.

Now test yourself

TESTED ☐

11 What type of cells make antibodies?
12 In which organ are lymphocytes made?
13 Define the term monoclonal antibodies.
14 Which techniques are used to collect monoclonal antibodies at the end of their manufacture?
15 Why are spleen cells fused with myeloma cells?
16 What diagnostic tests are monoclonal antibodies used for?
17 Which hormone do monoclonal antibodies bind with in the urine of pregnant women?
18 Why has the use of monoclonal antibodies not developed into a 'magic bullet' capable of treating many illnesses?
19 Define the term myeloma.
20 What part of a substance do monoclonal antibodies bind with?

Answers on p. 114

Plant disease

Detection and identification of plant diseases

REVISED ☐

Pathogens are micro-organisms that spread disease. These infect plants as well as animals including us. Pathogens can be viruses, bacteria and fungi. Plants can also be infected by insects.

(H) Plant diseases are often identified by their symptoms. These include:
- stunted growth
- spots on leaves
- areas of decay (rot)
- growths and **lesions**
- malformed stems or leaves
- discolouration
- the presence of pests.

Identification can be made by:
- Looking in a gardening manual (book) or website.
- Taking an infected plant to a laboratory.
- Using testing kits that contain monoclonal antibodies.

Viral diseases

The tobacco mosaic virus infects many plants including tomatoes, not just tobacco. Infected plants develop a mosaic-like pattern on their leaves. This reduces the amount of photosynthesis that can be completed which reduces growth. This disease costs an estimated 60 billion US dollars per year.

Fungal diseases

Black spot is a fungal pathogen which infects roses. It causes purple on black marks on the leaves of infected plants. As with the tobacco mosaic virus, this reduces the photosynthetic ability of the plant reducing its growth. Severely infected plants die. It is spread by water or wind. It can be treated by using fungicides and removing and destroying the affected leaves.

Other diseases

Aphids are small insects that are commonly found in large numbers on the stems of roses in our gardens. They suck the sap from plants which reduces growth, and eventually causes them to wilt. Aphids are eaten by ladybirds.

Plants can also suffer from ion **deficiency** diseases. These occur when they are unable to absorb enough mineral ions. Plants without suitable nitrate ions develop stunted growth because they cannot make enough protein. Magnesium ions are needed by plants to make chlorophyll, which is necessary for photosynthesis. Plants with insufficient magnesium appear yellow. We call this chlorosis.

Plant defence responses

REVISED

Just like you, plants have physical, chemical and mechanical methods of preventing infection by pathogens.

Physical defence responses include:
- strong cellulose cell walls that stop infection of individual cells
- tough waxy cuticle on leaves
- bark which forms layers of dead cells around stems which fall off taking the pathogens with them.

Chemical defence responses include:
- production, by a small number of plants, of antibacterial chemicals to stop infection (mint and witch hazel plants)
- the presence of poisons to stop leaves being eaten by herbivores (poison ivy and deadly nightshade plants).

Lesion: A part of an organism that has symptoms of disease, such as a wound, ulcer or abscess.

Deficiency: A lack or shortage.

Figure 3.5 **The tobacco mosaic virus damages the leaves of infected plants.**

Figure 3.6 **The effects of rose black spot are similar to the tobacco mosaic virus but are caused by a fungus.**

Mechanical adaptations include:
- thorns and hairs which stop leaves being eaten by herbivores
- leaves that curl up and droop when touched (the *Mimosa pudica* plant)
- spots on their leaves which mimic butterfly eggs to stop others laying eggs on them (the *Passiflora* plant).

Now test yourself

TESTED ☐

21 State two ways in which plant diseases are identified.
22 What are lesions?
23 State two ways in which diseases can be identified.
24 State an example of a viral infection of a plant and its symptoms.
25 State an example of a fungal infection of a plant and its symptoms.
26 How do aphids affect plants?
27 What symptoms would indicate that a plant is low in magnesium?
28 State an example of a physical defence in plants?
29 State an example of a chemical defence in plants?
30 State an example of a mechanical adaptation in plants?

Answers on p. 114

Summary

- Communicable diseases are those which are transmitted from person to person. Pathogens are viruses, bacteria, protists and fungi that cause communicable diseases to plants and animals. They are transmitted by direct contact, water or air.
- Bacteria reproduce inside the body and can produce toxins. Viruses reproduce inside cells.
- Measles is caused by a virus. Symptoms include a fever and red skin rash. HIV is a virus which initially causes flu-like symptoms. HIV develops into AIDS when a person's immune system cannot function anymore. Tobacco mosaic virus infects plants causing discoloured leaves which reduces growth.
- *Salmonella* food poisoning is spread in food prepared in unhygienic conditions. The symptoms include fever, abdominal cramps, vomiting and diarrhoea. Gonorrhoea is a bacterial STD which causes thick yellow or green discharge from the vagina or penis.
- Rose black spot is caused by a fungus. Infected plants have discoloured leaves which reduces growth. Malaria is transmitted by a protist carried within a mosquito. Symptoms include recurring fever and can be fatal.
- Non-specific defence systems against infection include skin, hairs within the nose, ciliated cells in the trachea and bronchi and stomach acid.
- If an invading pathogen passes these defences, the immune system attacks. White blood cells called phagocytes engulf and destroy pathogens. White blood cells called lymphocytes produce antibodies which 'clump' together pathogens. They also produce antitoxins to neutralise toxins produced by the pathogen. Lymphocytes 'remember' pathogens and are produced more quickly in subsequent exposures to them.
- Antibiotics are medicines which kill bacteria but not viruses. Their use has saved many lives. However, bacteria like methicillin-resistant *Staphylococcus aureus* (MRSA) are evolving to be antibiotic resistant. This is of great concern. Painkillers are used to treat symptoms but not kill pathogens.
- Traditionally drugs were extracted from plants and micro-organisms. Newer drugs are often synthesised by chemists. The discovery and development of new drugs is expensive and time-consuming. This tests toxicity, efficacy and dosage. Preclinical trials are completed on cells, tissues and live animals. Clinical trials then use healthy volunteers and patients.
- Ⓗ Monoclonal antibodies are produced from a single clone of cells. A mouse is injected with an antigen and naturally makes antibodies to this. Spleen cells are removed and fused with cancer cells called myeloma cells to form hybridoma cells. These are grown in large numbers and produce antibodies specific to the original antigen.

H
- Monoclonal antibodies are used in pregnancy test kits, to measure hormones and other chemicals in the blood, in research to identify specific molecules within cells and to treat some diseases like cancer.
- Monoclonal antibodies have produced more side effects than expected.
- Plant disease can be detected by stunted growth, spots on leaves, areas of decay (rot), abnormal growths, malformed stems and leaves, discolouration and the presence of pests.

H
- Identification can be made by using a gardening manual or website, taking infected plants to testing laboratories and using testing kits that contain monoclonal antibodies.
- Plants can be infected by viral, bacterial and fungal pathogens and insects.
- Plants can be damaged by nitrate deficiency which causes stunted growth, and magnesium deficiency which causes chlorosis.

Exam practice

1 What causes rose black spot? [1]
   A Viruses
   B Fungi
   C Bacteria
   D Protists
2 Who discovered the first antibiotic? [1]
   A Louis Pasteur
   B Charles Darwin
   C Alexander Fleming
   D Alfred Russel Wallace
3 Define the term antiseptic. [1]
4 What are the symptoms of the tobacco mosaic virus and what effect does this have upon the plant? [2]
5 What part of a pathogen do antibodies bind with? [1]
6 Label the two points on the graph shown by the lines. Describe the shape of this graph. [4]

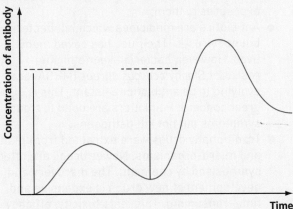

7 You have just completed an experiment into the effect of antibiotics or antiseptics on bacterial growth using agar plates. You have measured the zones of inhibition around the antibiotics and antiseptics. How can you tell which is the most effective drug? [1]
8 Describe how your non-specific defence systems stop you from falling ill. [6]
H 9 Describe the process of making monoclonal antibodies. [6]

Answers and quick quizzes online

4 Bioenergetics

Photosynthesis

Photosynthetic reaction

REVISED

Plants take in water through their roots and carbon dioxide through their leaves. These reactants are converted into glucose during photosynthesis. Oxygen is made as a by-product of this reaction. The word equation for this reaction is:

$$\text{carbon dioxide} + \text{water} \xrightarrow{\text{light in}} \text{glucose} + \text{oxygen}$$

The balanced symbol equation for photosynthesis is:

$$6CO_2 + 6H_2O \xrightarrow{\text{light in}} C_6H_{12}O_6 + 6O_2$$

Photosynthesis only occurs in light. It is an **endothermic reaction** so requires energy from light. Photosynthesis occurs in a green compound called chlorophyll which is present in sub-cellular structures called chloroplasts. These are found in high numbers in palisade mesophyll cells which are in turn found in the top layers of leaves.

> **Exam tip**
>
> In both the word and symbol equations, the two reactants and the two products can be either way around.

> **Endothermic reaction**: A reaction that requires heat to be absorbed to work.

Photosynthetic algae

Algae can photosynthesise as well as plants. Some algae are single-celled whilst others like seaweed are larger. Some algae have chlorophyll like plants, but others use different coloured photosynthetic pigments. These are seen in seaweeds which are green, brown and red. More than two-thirds of the oxygen made every day is by photosynthetic algae in our oceans and not plants on land.

> **Typical mistake**
>
> Plants complete photosynthesis to make glucose. Oxygen is a by-product. They do not make it for us.

Rate of photosynthesis

REVISED

The rate of reaction is how quickly it occurs. The rate at which plants and algae photosynthesise decreases when:
- temperatures fall (less kinetic energy present)
- carbon dioxide levels drop
- light intensities reduce (see below)
- plants do not have sufficient chlorophyll.

H If one of more of these conditions occurs, the rate of photosynthesis becomes limited. This or these are then '**limiting factors**'. It is important for farmers to reduce any limiting factors to ensure the maximum **yield** of their crops. They do this by:
- keeping plants warm in greenhouses or polytunnels
- keeping burners in greenhouses or polytunnels to produce carbon dioxide
- providing plants with maximum light levels.

> **Limiting factor**: Anything that reduces or stops the rate of a reaction.
>
> **Yield**: The amount of an agricultural product.

Required practical 6

Light intensity and photosynthesis

Aim: To investigate the effect of light intensity on the rate of photosynthesis using an aquatic organism such as pondweed

Equipment: Lamp, metre rulers, boiling tube, pondweed, water.

Method:
1 Set up the equipment as shown in Figure 4.1.
2 Place the boiling tube with pondweed 10 cm from the lamp.
3 Switch the light on and wait 2 minutes for the plant to acclimatise.
4 Record the number of bubbles of oxygen given off in 1 minute.
5 Move the boiling tube a further 10 cm from the lamp and repeat.
6 Repeat by increasing the distance every 10 cm up to 60 cm.

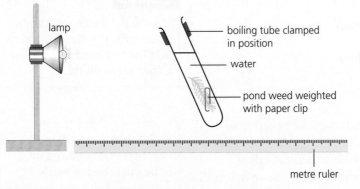

Figure 4.1 The equipment used to investigate the effects of light intensity on the rate of photosynthesis.

Results: Your results will look like those in Figure 4.2.

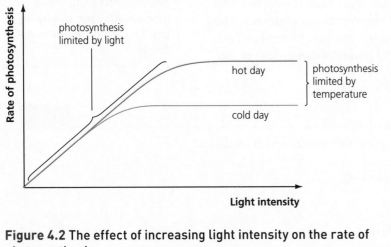

Figure 4.2 The effect of increasing light intensity on the rate of photosynthesis.

> **Exam tip**
>
> You should be able to describe the effects of temperature, light intensity carbon dioxide concentration and the amount of chlorophyll on the rate of photosynthesis. You should also be able to interpret graphs of photosynthesis rate involving one limiting factor.

> **H** **Exam tip**
>
> You should also be able to interpret graphs of photosynthesis rate involving more than one limiting factor. You should also be able to use the inverse square law and light intensity in the context of photosynthesis.

> **Exam tip**
>
> You should be able to use the method in Required practical 6 to measure and calculate the rate of photosynthesis.

Uses of glucose from photosynthesis

REVISED

Photosynthesis provides plants and algae with chemical energy in the form of glucose. They use this in five key ways shown in Figure 4.3.

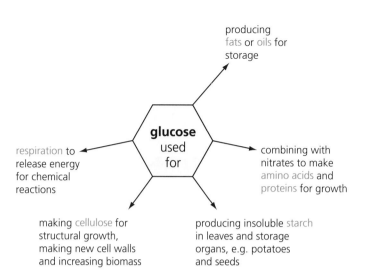

Figure 4.3 How plants make use of the glucose produced in photosynthesis.

Plants or algae doing photosynthesis are present at the bottoms of almost all food chains on our planet. So, photosynthesis provides the energy that supports almost all life on Earth including your own. Without photosynthesis there would be very little life on our planet.

Now test yourself

TESTED

1 State the word equation for photosynthesis.
2 State the balanced symbol equation for photosynthesis.
3 Besides plants, what other group of organisms completes photosynthesis?
4 Why is photosynthesis an endothermic reaction?
5 In what type of cells does most photosynthesis in leaves occur?
6 What four factors affect the rate of photosynthesis?
H▶ 7 How do farmers reduce limiting factors for photosynthesis?
8 Define the term yield.
9 State the equipment that you would use to investigate how light intensity affects the rate of photosynthesis.
10 Where are photosynthetic organisms usually found in a food chain?

Answers on p. 114

Respiration

Aerobic respiration

REVISED

Energy is released from glucose during **aerobic** respiration. This reaction is essential for many living organisms. It occurs continuously in specially adapted cell components called mitochondria which are found in the cytoplasm. The energy that is released is used to complete the seven life processes: movement, reproduction, respiration, sensitivity, nutrition, excretion and growth. The word equation for this reaction is:

glucose + oxygen $\xrightarrow{\text{energy out}}$ carbon dioxide + water

The balanced symbol equation for respiration is:

$C_6H_{12}O_6 + 6O_2 \xrightarrow{\text{energy out}} 6CO_2 + 6H_2O$

> **Aerobic:** In the presence of oxygen.

> **Exam tip**
>
> In both the word and symbol equations, the two reactants and the two products can be either way around.

Respiration is an **exothermic reaction**. It transfers energy into its surroundings. Unlike photosynthesis, respiration occurs at all times not just during the day.

> **Exothermic reaction**: A reaction that gives out heat energy.

Conversion of energy in respiration (and photosynthesis)

REVISED

When the equations for photosynthesis and respiration are looked at, they initially seem to be the opposite of each other.

The reactants in respiration are the same as the products in photosynthesis and the other way around. However, there is one crucial difference. Photosynthesis is endothermic and so takes energy from its surroundings (light) and stores this in glucose. Respiration is exothermic and so releases energy from glucose.

The equations for photosynthesis and respiration work together for almost all life on our planet. The Sun's energy drives almost all food chains. Energy transferred by light (mainly from the Sun) is converted by plants and algae into a chemical store of energy called glucose during photosynthesis. Energy for life processes is then transferred from glucose during respiration. This energy has two main uses. It is converted into:

1 Heat energy to keep **homoeothermic** (warm-blooded) animals warm.
2 A chemical store of energy that is used in reactions to build larger molecules and processes like movement.

> **Exam tip**
> You should also be able to describe cellular respiration as an exothermic reaction which is continuously occurring in all cells.

> **Homoeothermic**: Warm blooded, like birds and mammals.

Anaerobic respiration

REVISED

Anaerobic respiration occurs when there is not enough oxygen to complete anaerobic respiration. Times like this occur when you have been exercising vigorously. You are unable to breathe quickly enough or deep enough to supply your cells with sufficient oxygen. This is the equation for anaerobic respiration:

$$\text{glucose} \xrightarrow{\text{energy out (only 5\%)}} \text{lactic acid}$$

When your cells respire anaerobically, there is insufficient oxygen to make carbon dioxide and water. So, an intermediary **product** called lactic acid is made instead. Many scientists think that lactic acid build-up in muscles causes cramp. During long periods of vigorous exercise, your muscles become fatigued and stop contracting efficiently.

Crucially, only 5% of the energy released in aerobic respiration is transferred during anaerobic respiration. The remaining 95% is stored within the lactic acid.

When you have finished exercising, your breathing rate and volume (depth) do not reduce to normal immediately. They remain high until your body has absorbed enough oxygen to pay your body back the oxygen you owe it. This is paying your **oxygen debt**. When this happens, the lactic acid can be fully broken down into carbon dioxide and water. This releases the remaining 95% of the energy that was originally stored in glucose. This process is shown in the following equation:

$$\text{lactic acid} + \text{oxygen} \xrightarrow{\text{energy out (only 95\%)}} \text{carbon dioxide} + \text{water}$$

> **Anaerobic**: In the absence of oxygen.
>
> **Product**: The substance or substances produced in a reaction.
>
> **Oxygen debt**: The temporary shortage of oxygen in respiring tissues and organs.

> **Exam tip**
> You should also be able to compare aerobic and anaerobic respiration.

(H) Lactic builds up in cells that are respiring anaerobically. This diffuses into the bloodstream from high concentration in the cells to a lower concentration in the blood. This is transported to the liver. Here it diffuses from a high concentration in the blood to a lower concentration in the liver. It is then converted back to glucose.

Anaerobic respiration in plants and micro-organisms

REVISED

All cells in your body must complete respiration to release the energy they need to live. The same is true of all plants and micro-organisms as well. Some micro-organisms such as the yeast fungus respire anaerobically too:

glucose $\xrightarrow{\text{energy out}}$ ethanol + carbon dioxide

This reaction in yeast is called **fermentation**. This is an economically important reaction for we use it to make bread and beer. Ethanol is commonly called alcohol. It is present in about 4% in beer, 12% in wine and 40% in spirits. Bread is not alcoholic because it is baked before we eat it. This evaporates away any alcohol.

Fermentation: The chemical break-down of glucose into ethanol and carbon dioxide by respiring micro-organisms such as yeast.

Metabolism

REVISED

Metabolism is the sum of all the chemical reactions in your body. These include the digestion of food, aerobic and anaerobic respiration and protein synthesis. Your metabolism is regulated by your thyroid gland. These chemical reactions either break down large molecules into smaller ones or the reverse.

Break-down reactions

Break-down reactions turn larger, often more complicated molecules into smaller ones. These reactions do not often require much energy from respiration. Any excess proteins you obtain from your diet are broken down to form amino acids in your liver. These in turn are broken down by removal of an amine group. This forms urea which is excreted by your kidneys in your urine.

Synthesis reactions

Synthesis reactions turn smaller, often less complicated molecules into larger ones. These reactions often do require energy from respiration. The following are examples of synthesis reactions:
- Glucose is produced by plants and algae during photosynthesis. They can store the glucose as insoluble starch or convert it to cellulose to make cell walls. In animals, excess glucose can be stored as **glycogen** which is stored in the liver.
- Lipids are fats (solids) and oils (liquids). These are made from three molecules of fatty acid and one of glycerol. Lipids are needed for cell membranes and as a store of chemical energy (fat).
- Glucose and nitrate ions are needed to make amino acids. These are joined together in the correct sequence to make proteins during protein synthesis.

Exam tip
You should also be able to explain the importance of sugars, amino acids, fatty acids and glycerol in the synthesis and break-down of carbohydrates, proteins and lipids.

Glycogen: An insoluble store of glucose in the liver.

Now test yourself

11 State the word equation for respiration.
12 State the symbol equation for respiration.
13 Why is respiration an exothermic reaction?
14 When do plants complete respiration and photosynthesis?
15 What are the two main uses of energy produced in respiration in warm-blooded animals?
16 What does anaerobic mean?
17 State the word equation for anaerobic respiration.
H 18 What is lactic acid broken down into?
19 State the word equation for fermentation.
20 What two types of reactions make up metabolism?

Answers on p. 114

Summary

- The word equation for photosynthesis is:

carbon dioxide + water $\xrightarrow{\text{light in}}$ glucose + oxygen

- The symbol equation for photosynthesis is:

$6CO_2 + 6H_2O \xrightarrow{\text{light in}} C_6H_{12}O_6 + 6O_2$

- Photosynthesis is an endothermic reaction in which energy is transferred from the environment to the chloroplasts in plant cells by light.
- The rate of photosynthesis is affected by temperature, light intensity, carbon dioxide concentration and the amount of chlorophyll.
- These limiting factors can interact. They are important in the economics of obtaining optimum conditions in greenhouses for growing food.
- Glucose produced in photosynthesis is used for respiration, converted into insoluble starch for storage, used to produce fats or oils for storage, used to produce cellulose for growth and amino acids for protein synthesis.
- Respiration can occur in the presence of oxygen (aerobic) or absence (anaerobic).
- The energy transferred by respiration supplies all the energy needed for living organisms. This is needed for chemical reactions to build larger molecules, movement and keeping warm (in warm-blooded animals).
- The equation for aerobic respiration is:

glucose + oxygen $\xrightarrow{\text{energy out}}$ carbon dioxide + water

- The balanced symbol equation for aerobic respiration is:

$C_6H_{12}O_6 + 6O_2 \xrightarrow{\text{energy out}} 6CO_2 + 6H_2O$

- During exercise, the heart and breathing rates increase together with the breath depth. This supplies respiring cells with more oxygenated blood. Without sufficient oxygen, anaerobic respiration occurs.
- The equation for anaerobic respiration is:

glucose $\xrightarrow{\text{energy out (only 5\%)}}$ lactic acid

- Only around 5% of the energy is released in anaerobic respiration. The rest remains within lactic acid. Oxygen debt is the amount of extra oxygen that is needed to react with and remove lactic acid. When an oxygen debt has been paid, lactic acid is broken down into carbon dioxide and water and the rest of the energy is released.
- H Lactic acid is transported to the liver where it is converted back to glucose.
- The equation for anaerobic respiration in plant and yeast cells is:

glucose $\xrightarrow{\text{energy out}}$ ethanol + carbon dioxide

- Anaerobic respiration in yeast is called fermentation. This is an economically important reaction in the making of bread and alcoholic drinks.
- Metabolism is the sum of all the reactions in a cell or organism. It includes converting glucose to starch, glycogen and cellulose. It includes forming lipids from fatty acids and glycerol. It includes using glucose and nitrate ions to form amino acids to make proteins. It also includes respiration and break-down of urea.

Exam practice

1 In what cell component does photosynthesis occur?   [1]
  A Chlorophyll
  B Ribosomes
  C Mitochondria
  D Chloroplasts

2 What is the product of anaerobic respiration (fermentation) in micro-organisms?   [1]
  A Lactic acid
  B Ethanol and carbon dioxide
  C Water and carbon dioxide
  D Glucose and oxygen

3 Define the term metabolism.   [1]

4 State what happens to excess amino acids in your blood. What type of reaction is this?   [2]

5 Why is photosynthesis important for all life on Earth not just plants and algae.   [2]

H► 6 State the word (standard tier) or symbol equations for photosynthesis.   [2]

7 Describe the method you would use to investigate the effects of light intensity on photosynthesis.   [4]

8 Describe the uses of glucose from photosynthesis.   [6]

9 Compare and contrast the reactions of aerobic respiration and photosynthesis.   [6]

Answers and quick quizzes online

ONLINE

5 Homeostasis and response

The human nervous system

Homeostasis

The millions of cells in your body have optimum conditions. They require glucose for respiration, water and a sufficient temperature to function. **Homeostasis** is the maintenance of these and many other conditions. We call it the maintenance of a constant internal environment. These changes happen without you knowing. They are automatic or involuntary.

> **Homeostasis:** The maintenance of a constant internal environment.

> **Exam tip**
>
> You should also be able to describe homeostasis as the regulation of the internal conditions of a cell or organism to maintain optimum conditions.

Structure and function

REVISED

Your nervous system controls your voluntary and involuntary actions. It allows you to react to your surroundings and coordinate your behaviour. It is made from millions of nerve cells which transmit and receive millions of messages each day. Your nervous system is made up from:

- Your **central nervous system (CNS)** – your brain and spinal cord
- Your peripheral nervous system – the millions of nerves that criss-cross the rest of your body.

Nerves are made from bundles of individual neurones.

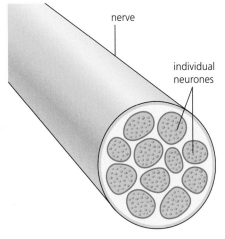

Figure 5.1 **A bundle of neurones is a nerve.**

> **Exam tip**
>
> You should also be able to explain how the structure of the nervous system is adapted to its functions.

> **Central nervous system (CNS):** The brain and spinal cord.
>
> **Receptor:** A cell or group of cells at the beginning of a pathway of neurones that detects a change and generates an electrical impulse.

Sending electrical impulses

All messages sent along neurones in your nervous system are electrical. These messages travel very quickly. The electrical signal is generated by a specific type of cell called a **receptor**. You have these cells in all your sense organs including your skin. Some areas of your skin like your lips and finger tips are very sensitive. Here there are more receptors than in less sensitive areas like your elbow.

Table 5.1 Your senses, the organs and stimuli involved.

Sense	Organ	Stimuli
Sight	Eyes	Light
Hearing	Ears	Sound
Taste	Tongue	Chemicals in food
Smell	Nose	Chemicals in air
Touch	Skin	Touch, pressure, temperature, pain and itch

Sensory, relay and motor neurones

There are three main types of neurone. Those that carry signals from receptors towards your central nervous system are called **sensory neurones**. Those that carry signals around your brain and spinal cord are called **relay neurones**. Those that carry signals away from your central nervous system are called **motor neurones**.

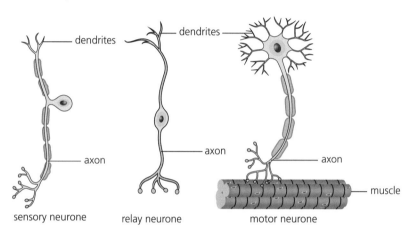

Figure 5.2 The three types of neurone.

Motor neurones end in muscles or glands. Muscles contract or relax to move parts of your body. Glands release hormones into the blood. We call both muscles and glands **effectors**. The pathway of an electrical signal from receptor to effector is:

stimulus → receptor → sensory neurones → relay neurones → motor neurones → effector → response

Synapses

There is more than one nerve that links each of your fingers to your brain. These nerves form a network. If one nerve is damaged, the signal can be rerouted and still reach your brain. If each electrical signal passes along multiple nerves, there must be gaps between them. These are called **synapses**.

When an electrical signal reaches the end of a neurone, it is quickly converted into a chemical one to cross the synapse. The ends of neurones are branched. At the tips of these branches, are special areas that make **neurotransmitters**. These chemicals diffuse across the synapse, bind to receptors and restart the electrical signal in the **dendrites** of the next nerve. This happens extremely quickly.

Revision activity

Draw out this table with only the headings along the top and the first column on the left. Try to fill in the rest of the table from memory to help you to revise.

Sensory neurone: A neurone that carries an electrical impulse from a receptor towards the central nervous system.

Relay neurone: A neurone that carries an electrical impulse within the central nervous system.

Motor neurone: A neurone that carries an electrical impulse away from the central nervous system to an effector (muscle or gland).

Effector: A muscle or a gland.

Synapse: A gap between the axon of one nerve and the dendrites of another where neurotransmitters transmit the impulse.

Neurotransmitter: A chemical substance released at the end of one neurone that diffuses across a synapse to begin a second electrical impulse in another neurone.

Dendrites: The branched beginnings of neurones, which can detect neurotransmitters and start another electrical impulse.

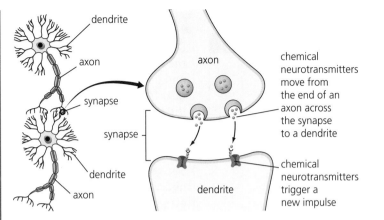

Figure 5.3 **How signals move from one nerve cell to another across a synapse.**

Reflex arc: The movement of an electrical impulse that avoids the brain to save time and so prevent damage to your body.
Reflex response: An automatic response that you do not think about.
Concordant results: Results that are similar.

The reflex arc

Some of your reactions are automatic. You do not think about moving your hand if you put it on a hot radiator. You just move it. These are called **reflex responses**. Here the electrical signal is not initially transferred to the conscious region of your brain. The signal is started by receptors and travels along sensory neurones (like normal) to your spinal cord. Here relay neurones immediately send a signal down motor neurones to your muscles, which contract and move your body. The signal is sent to your conscious brain shortly afterwards. This makes your reactions quicker which potentially reduces damage to your body.

Exam tip
You should be able to explain how the structure of synapses, and sensory, relay and motor neurones all relate to their function. You should also be able to explain the importance of reflex actions.

Required practical 7

Reaction times

Aim: To plan and carry out an investigation into the effect of a factor on human reaction time.

Equipment: Ruler

Method:
You will need a partner to complete this investigation.

1 Person A should hold the end of a ruler between the finger and thumb of person B. Without warning, person A should drop the ruler for person B to catch.

2 The shorter the distance the ruler drops, the faster the reactions of person B.

3 Charts that convert distance to reaction time are easily found on the internet.

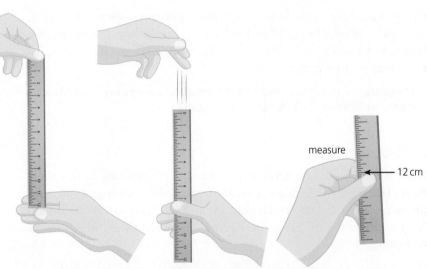

Figure 5.4 **How to test your partner's reaction time.**

4 Repeat this until you have three **concordant results**, ignore anomalous results and calculate the mean.

5 Drink a can of caffeinated drink and wait a short period for it to be absorbed into your bloodstream.

6 Repeat and compare your results.

Results: Most people find that their reaction time reduces after drinking a caffeinated drink. This means their reactions are faster.

The brain

Structure of the brain

Your brain controls all your body's complex behaviour. It is made from billions of interconnected relay neurones. It has different regions that complete different functions. These are shown in Figures 5.5 and 5.6.

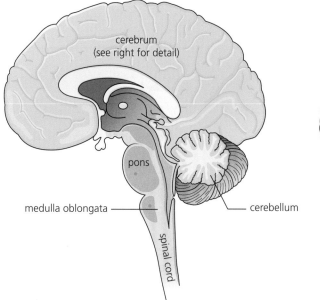

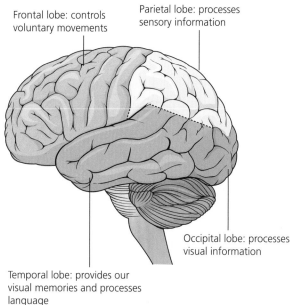

Frontal lobe: controls voluntary movements

Parietal lobe: processes sensory information

Occipital lobe: processes visual information

Temporal lobe: provides our visual memories and processes language

cerebrum (see right for detail)

pons

medulla oblongata

cerebellum

spinal cord

Figure 5.5 The main regions of the brain (cut away to show inside).

Figure 5.6 The functions of the parts of the cerebrum (not cut away).

> **Medulla oblongata**: A part of your brain above the spinal cord that controls your breathing and heart rates.
>
> **Cerebellum**: A part of the back of the brain that coordinates muscular activity.
>
> **Cerebrum**: The outer layer of the brain, which plays an important role in consciousness.
>
> **Pons**: A part of the brainstem which links the cerebrum to the medulla oblongata and is responsible for sleep, respiration, swallowing and a number of other factors.

ⓗ Studying the brain

Neuroscience is the study of the brain. Much of the brain looks very similar and so detecting different regions and their functions was initially very difficult. A French surgeon called Paul Broca (1824–80) helped to develop our understanding of neuroscience. A small region of the frontal lobe of the brain is called the **Broca's area** after his work. He found that patients with damage to this part of the brain had difficulty speaking, and so linked this area to speech.

More recent medical research into the brain has also involved patients with brain damage. They have had parts of their brains stimulated using electricity and complex equipment, like magnetic resonance imaging (MRI) scanners, are used to see the results.

The brain is a very complex and delicate organ. This makes its study and the treatment of brain disorders particularly challenging.

> **Broca's area**: A part of the frontal lobe, named after Paul Broca, that controls speech.

The eye

The eye is the organ of vision and it allows you to see. It does this by detecting light and converting it into electrical signals that are passed along nerves to your brain.

Structure of the eye

The structure of the eye is shown in Figure 5.7 and the functions of its components are given in Table 5.2.

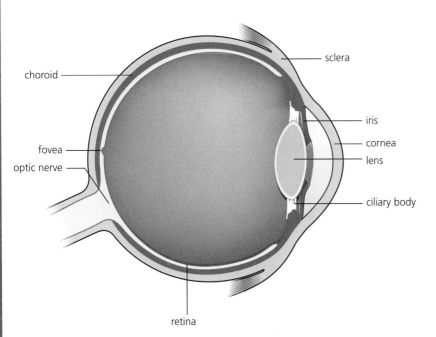

Figure 5.7 The structure of the mammalian eye in cross-section.

> **Exam tip**
>
> You should be able to identify the positions of the structures of the eye and explain how they relate to their functions.

> **Revision activity**
>
> Draw out Table 5.2 with only the headings along the top and the first column on the left. Try to fill in the rest of the table from memory to help you to revise.

> **Rods**: Light-sensitive receptor cells on your retina in the back of your eye that let you see in low light conditions.
>
> **Cones**: Light-sensitive receptor cells on your retina in the back of your eye that let you see in colour.

Table 5.2 The parts of the mammalian eye and their functions.

Part	Function
Cornea	The transparent part of your eye that covers the iris and pupil. It refracts light through the pupil.
Lens	The biconvex structure that sits behind your pupil. It further refracts light to be focused on your retina. Its shape is changed in a process called accommodation when you look at near and far objects.
Iris	The coloured muscle that surrounds your pupil. It controls the size of the pupil by relaxing and contracting to let more or less light in.
Ciliary body	A structure made of muscles called ciliary muscles, which can contract and relax to change the shape of your lens during accommodation.
Choroid	The layer of your eye found between the retina and the sclera. It provides oxygen and nourishment to the cells of the retina.
Retina	The layer of receptors found inside your eye. It contains two types of light-sensitive cells called **rods** and **cones**.
Fovea	A specific part of the retina that is responsible for your sharpest vision. About half of the neurones that leave your eye come from this point alone.
Sclera	The white of your eye. It is the outer layer that protects the rest of your eye. In many other mammals it is not white and so is not as easy to distinguish from the iris.
Optic nerve	The nerve made from neurones that connect your eyes to your brain. Electrical impulses pass from the eyes to the brain to let you see.

Accommodation

The lenses in your eyes can change shape to allow you to see near or far objects. This process is called the **accommodation** reflex and is automatic. The changes in lens shape during accommodation are shown in Figure 5.8.

To see a near object, the muscles in your ciliary body contract, the suspensory ligaments loosen and the lens becomes thicker. This refracts light rays strongly. To see a distant object, your ciliary muscles relax, the suspensory ligaments tighten and the lens becomes thinner. This refracts light rays less.

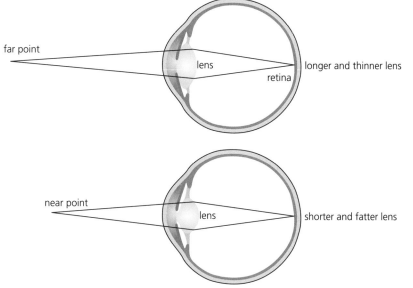

Figure 5.8 When you look at near objects your lens is short and fat in shape. When you look at far objects your lens is longer and thinner.

Long-sightedness and short-sightedness

Some people are born unable to correctly focus light onto their retina. People who cannot see objects that are close are called long-sighted. This is also called **hyperopia**. Those that cannot see distant objects are short-sighted (**myopia**). Both hyperopia and myopia are treated by eyeglasses or contact lenses. Both treatments help by refracting light before it reaches the lens. More recent laser eye treatment can now permanently change the shape of the cornea so people no longer need to wear eyeglasses or contact lenses. In addition, replacement lenses can now be inserted into a patient's eyes.

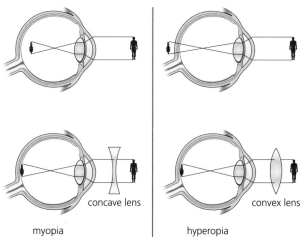

Figure 5.9 Myopia and hyperopia can be treated by wearing different types of glasses.

Accommodation: The changing shape of the lens in your eye to focus on near or far objects.

Hyperopia: A medical condition called long-sightedness in which people cannot clearly see objects close to them.

Myopia: A medical condition called short sightedness in which people cannot clearly see objects far away.

> **Exam tip**
>
> You should be able to interpret ray diagrams showing long-sightedness and short-sightedness and explain how lenses correct them.

Control of body temperature

Your body has evolved to function most effectively between 36.5 and 37.5 °C. This is the optimum temperature at which most of your enzymes will function. The thermoregulatory centre in your brain controls your body temperature. It contains receptors which detect the temperature of the blood. Your skin also contains temperature receptors which detect changes in temperature and begin electrical signals that travel along nerves to your thermoregulatory centre.

Sweating and shivering

If your thermoregulatory centre detects that you are too hot, sweat glands in your skin produce sweat. This evaporates from your skin transferring away energy to cool you.

When your body becomes too cold you automatically begin to shiver. This is a reflex response which you cannot control. Your skeletal muscles contract and relax very quickly which causes the shivers. This generates heat to warm you.

Vasoconstriction and vasodilation

When your body becomes too hot, it opens blood capillaries near the surface of your skin. This sends more warm blood there to lose excess heat. This is called **vasodilation**. On hot days your cheeks appear flushed. This is the increase of blood towards the surface of your skin.

The reverse happens if your thermoregulatory centre detects you are too cold. The same blood capillaries close to reduce the volume of blood near your skin that can lose heat. This is called **vasoconstriction**. On cold days your skin is often paler. This is the reduction of blood towards the surface of your skin.

(a)

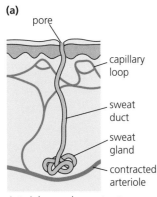

pore

capillary loop

sweat duct

sweat gland

contracted arteriole

Arteriole muscles contract to reduce blood flow through skin capillaries.

(b)

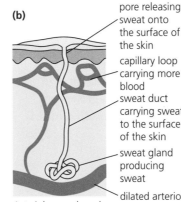

pore releasing sweat onto the surface of the skin

capillary loop carrying more blood

sweat duct carrying sweat to the surface of the skin

sweat gland producing sweat

dilated arteriole

Arteriole muscles relax to allow more blood flow through skin capillaries.

Figure 5.10 Cross-sections of the skin on (a) a cold day, and (b) a hot day.

Hypothermia and hyperthermia

A body temperature increase from between 37.5 to 38.5 °C is called a fever. Above 38.5 °C **hyperthermia** begins. This is often because of heat stroke or an allergic reaction. Without a reduction in temperature, hyperthermia can be fatal.

A body temperature decrease below 35 °C is called **hypothermia**. This often results in mental confusion. People with hypothermia may

Vasodilation: The increase in size of blood vessels to increase the flow of blood to the surface of the skin and therefore increase heat loss.

Vasoconstriction: The reduction in size of blood vessels to reduce the flow of blood to the surface of the skin and therefore reduce heat loss.

Hyperthermia: A medical condition caused by high body temperature, which causes organ failure and eventually death.

Hypothermia: A medical condition caused by low body temperature, which causes organ failure and eventually death.

actually try to take their clothes off. Without an increase in temperature, hypothermia is fatal.

Now test yourself

TESTED ☐

1 Define the term homeostasis.
2 What are effectors?
3 Describe the pathway of an electrical signal from receptor to effector.
4 How is the pathway of a reflex action different from a 'normal' nervous response?
5 State the function of the medulla oblongata and its location within the brain.
H▶ 6 Why is it difficult to study the brain?
7 What two types of cells are found in the retina and what are their functions?
8 How does the lens change shape during accommodation when viewing a far-away object?
9 What occurs during vasodilation?
10 What medical condition occurs when a person's body becomes too hot?

Answers on pp. 114–115

Hormonal control in humans

Human endocrine system

REVISED ☐

Your **endocrine system** is a group of glands that secrete hormones into your blood. These glands, their hormones and functions are given in Table 5.3 and shown in Figure 5.11. Hormones travel in the blood and so hormonal responses are usually slower than electrical signals that are carried by the nerves of the nervous system.

> **Endocrine system**: The system of glands that secrete hormones into the circulatory system.

Table 5.3 **Common examples of hormones and their functions.**

Hormone	Produced	Target organ	Function
ADH (anti-diuretic hormone)	Pituitary gland	Kidney	Controls the concentration of water in urine
TSH (thyroid-stimulating hormone)	Pituitary gland	Thyroid	Controls the release of hormones from your thyroid gland
Adrenaline	Adrenal gland	Heart (and other vital organs)	Prepares the body to fight or run away (flight)
Insulin and glucagon	Pancreas	Liver	Insulin increases and glucagon decreases the conversion of blood glucose to glycogen
Thyroid hormones (e.g. thyroxine)	Thyroid	Various	Control how quickly you use energy, make proteins and how sensitive your organs are to other hormones
Oestrogen	Ovaries	Reproductive organs	Controls puberty and the menstrual cycle in women
Testosterone	Testes	Reproductive organs	Controls puberty in men

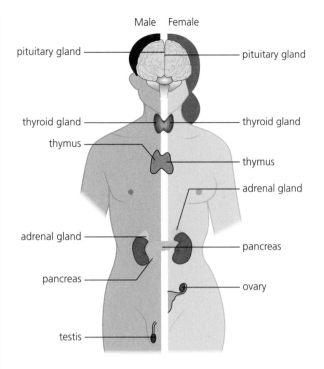

Male   Female

pituitary gland — ... — pituitary gland

thyroid gland — ... — thyroid gland

thymus

— thymus

— adrenal gland

adrenal gland —

— pancreas

pancreas —

— ovary

testis —

Figure 5.11 The positions of key glands in your endocrine system.

The **pituitary gland** in your brain is your 'master gland'. It secretes hormones that directly control growth and blood pressure, and partly controls how your kidneys, and ovaries or testes function. It also partly controls pregnancy and childbirth. It does this by releasing hormones that control other glands.

Pituitary gland: A gland in your brain that produces growth hormones, anti-diuretic hormone (ADH), thyroid-stimulating hormone (TSH), follicle-stimulating hormone (FSH) (in women) and luteinising hormone (LH) (again in women).

Exam tip

You should also be able to describe the principles of hormonal coordination and control by the endocrine system. You should also be able to identify the position of the glands in Figure 5.11.

Control of blood glucose concentration

REVISED

Your digestive system breaks down food into smaller, soluble molecules that are absorbed into your blood. Carbohydrase enzymes break down carbohydrates like starch into sugars like glucose. Many people eat three times a day but all your cells need glucose all the time to release energy during **respiration**. Your body absorbs excess sugar and stores it for times when you need it. This is an example of homeostasis. The concentration of your blood glucose is monitored and controlled by your pancreas.

Insulin and glucagon

When your blood glucose level is too high, your pancreas releases the hormone **insulin**. This travels in the bloodstream to your liver and muscle cells which convert excess glucose into insoluble glycogen. This reduces your blood glucose levels and returns them to normal.

When your blood glucose level is too low, your pancreas releases the hormone **glucagon**. This travels in the bloodstream to the liver which converts glycogen back to soluble glucose. This is released into your blood so it increases your blood glucose levels and returns them to normal.

Respiration: The release of energy from glucose.

Insulin: A hormone produced in your pancreas that lowers blood glucose by converting it to glycogen and storing it in the liver.

Glucagon: A hormone produced in the pancreas that raises blood glucose by breaking down glycogen stored in the liver.

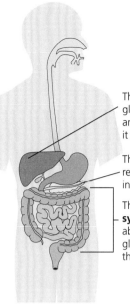

The **liver** stores glucose as glycogen and releases it when it is needed.

The **pancreas** releases the hormones insulin and glucagon.

The **digestive system** absorbs glucose into the blood.

Figure 5.12 Absorption of glucose, control of blood glucose concentration and storage of glucose in the body.

This is an example of **negative feedback control**. This occurs when your body detects a change (too little or too much blood glucose), makes a change and returns itself to normal. This change is therefore an example of homeostasis.

Diabetes

The causes of **type 1 diabetes** are unknown. It develops in children or young adults. It results in a person's immune system mistakenly destroying the insulin-producing cells in their pancreas. This means they cannot produce insulin. There is no cure. People with type 1 diabetes usually inject insulin, eat carefully and exercise regularly to control their blood glucose concentration.

Type 2 diabetes usually develops later in life than type 1. People with type 2 diabetes cannot produce enough insulin or if they can their liver and muscle cells do not respond to it. They do not absorb excess glucose and store it as glycogen. People with type 2 diabetes often feel thirsty, urinate more frequently and feel tired. Type 2 diabetes develops more commonly in people who do not exercise regularly, have a high sugar diet and are obese. There is no cure for type 2 diabetes either. Insulin is not injected (because their cells do not respond to it) but people with type 2 diabetes eat carefully and exercise regularly.

Maintaining water and nitrogen balance in the body

REVISED

The cytoplasm of all your cells is made primarily of water. It is essential for life. Other organisms like camels can go without water for longer periods of time, but all life on Earth eventually needs water to survive. We are looking for water in other parts of our Solar System to find extra-terrestrial life.

Exam tip

You should also be able to explain how insulin controls blood glucose levels.

Typical mistake

It is important you do not mistake the hormone glucagon for the storage molecule glycogen.

H **Exam tip**

You should also be able to explain how glucagon interacts with insulin in a negative feedback cycle to control blood glucose levels.

Negative feedback control: A homeostatic mechanism by which the body detects a change and makes an adjustment to return itself to normal.

Type 1 diabetes: A medical condition that usually develops in younger people, preventing the production of insulin.

Type 2 diabetes: A medical condition that usually develops in later life, preventing the absorption of insulin.

Exam tip

You should also be able to compare type 1 and type 2 diabetes and explain how they are treated. You should be able to explain the differences between graphs of blood glucose levels of people with and without diabetes.

Water balance

Water is lost from your body when you breathe out, sweat (which also contains ions and urea) and urinate. These actions happen automatically. You cannot control water loss by your lungs or skin. Your body can control water loss by determining how much urine you pass and the concentration of water in it. On hot days when you are dehydrated your urine is dark yellow in colour because it is more concentrated. Your body is retaining more water.

Your kidneys control how much water, urea and ions are released. They do this by filtering these substances from your blood and retaining that which you need. The rest are released in your urine, which is released from your kidneys and stored in your bladder.

ⓗControl of water balance

Your water concentration is monitored by the osmoregulatory centre of your brain. This gland releases **anti-diuretic hormone (ADH)** into your blood which is transported to the kidneys. Too much water in your body means less ADH is produced which means you urinate more, dilute urine. Too little water in your body means more ADH is produced. This means you will pass a smaller volume of more concentrated urine.

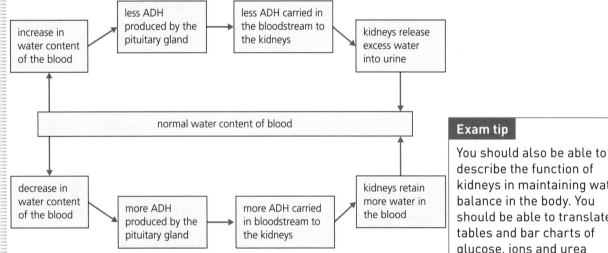

Figure 5.13 A flow diagram summarising the control of water content in the blood by ADH and the kidneys.

This is an example of negative feedback control. This occurs when your body detects a change (too little or much water), makes a change and returns itself to normal. This is therefore an example of homeostasis.

ⓗProtein removal

Excess proteins from your diet are broken down by your digestive system to form amino acids. In your liver, amino acids are broken down by a process called deamination. This forms ammonia which is toxic and so is immediately converted to urea. This is excreted by your kidneys in your urine.

The kidney

You have two kidneys, one either side of your spine in your lower back. They are supplied with blood by your renal arteries. Within each of your kidneys are about a million smaller structures called **nephrons**. These

> **Exam tip**
>
> You should also be able to describe the function of kidneys in maintaining water balance in the body. You should be able to translate tables and bar charts of glucose, ions and urea before and after filtration.

> ⓗ **Exam tip**
>
> You should also be able to describe the effect of ADH in maintaining water balance in the body.

> **Anti-diuretic hormone (ADH)**: A hormone produced in your pituitary gland that regulates the volume of urine you excrete.
>
> **Nephron**: The functional unit of the kidney where excess water, ions and urea are removed from the blood.

remove excess water, ions and urea from your blood before returning it in your renal veins.

The process in your kidneys has three steps:
1 All glucose, urea, ions and water are removed from the blood (filtration).
2 All glucose, and some ions and water are selectively reabsorbed back into the blood (selective reabsorption).
3 All urea, and excess ions and water are excreted in your urine (excretion).

So all glucose is retained and all urea is removed. Appropriate ions and water are retained to maintain optimum levels.

Kidney failure

Injuries, communicable diseases and inherited disorders can all reduce the function of kidneys. This is called kidney failure and can be fatal if not treated. A permanent solution to kidney failure is a transplant. Because we have two kidneys and can survive with only one, kidney transplants can come from living donors as well as recently dead ones. A kidney transplant can only come from a suitable donor. The body of the recipient will recognise the kidney as foreign and attack it unless it comes from a closely related person. This is often the closest family member. Patients often take suppressing drugs for the rest of their lives to stop their body rejecting the donated kidney.

Whilst patients are waiting on donor lists for a suitably matched kidney they will undergo **kidney dialysis**. Their blood is temporarily removed from their body by a machine which filters all urea from it before returning it to them. This takes several hours and must happen several times each week. This is not a permanent cure.

> **Kidney dialysis:** A temporary medical treatment for kidney failure patients in which blood is removed from their body, filtered and then replaced.

Hormones in human reproduction

REVISED

The sex hormone in men is **testosterone**, which is secreted from the testes. In women, it is the hormones **oestrogen** and **progesterone**, which are produced by the ovaries. These hormones are responsible for puberty in both sexes. Oestrogen and progesterone also regulate the menstrual cycle.

The menstrual cycle

From puberty to the **menopause**, at between 45 and 55 years old, women undergo a 28-day reproductive cycle called the menstrual cycle. Unless a woman is pregnant, the menstrual cycle begins on day one with the break-down of the lining of the uterus from the previous cycle. This process is called **menstruation** or having a period and lasts for a few days. It can be painful and cause cramps. Shortly after this process, the hormone oestrogen is produced by the ovaries and causes the lining of the uterus to start to thicken again. It is preparing itself for a fertilised ovum (egg) to settle and for the woman to become pregnant.

Around day 14 of the cycle, an ovum is released from an ovary. This is called ovulation. In the days following ovulation a woman is at her most fertile. If sperm are ejaculated into her vagina, they can swim upwards past the cervix and through the uterus to meet the ovum in a fallopian tube. If the sperm fertilises the ovum, it is likely to settle into

> **Testosterone:** A male sex hormone produced in the testes that controls puberty.
>
> **Oestrogen:** A female sex hormone produced in the ovaries that controls puberty and prepares the uterus for pregnancy.
>
> **Progesterone:** A female sex hormone produced in the corpus luteum that prepares the uterus for pregnancy.
>
> **Menopause:** The point in a woman's life, usually between 45 and 55, when she stops menstruating after which she cannot become pregnant.
>
> **Menstruation:** Having a period, as a part of the menstrual cycle.

the lining of the uterus and develop into a baby. If this happens, levels of the hormone progesterone stay high in the woman's body. This stops her from having another period, which would result in a natural abortion. If the ovum is not fertilised, then progesterone levels drop towards the end of the cycle and the woman menstruates again.

There are four hormones that work together in the menstrual cycle. They are described in Table 5.4 and their effects are shown in Figure 5.14.

> **Follicle-stimulating hormone (FSH):** A hormone produced by the pituitary gland that causes an ovum to mature in an ovary and stimulates the ovary to produce oestrogen.

Table 5.4 The main hormones involved in the control of the menstrual cycle.

Hormone	Released by	Target organ and effect
Follicle-stimulating hormone (FSH)	Pituitary gland	Ovary ● Causes an ovum to mature in the ovary inside the **follicle** ● Stimulates ovaries to produce oestrogen
Oestrogen	Ovaries	Uterus ● Causes lining to thicken in first half of the cycle ● High oestrogen concentration switches off the release of FSH and switches on the release of LH
Luteinising hormone (LH)	Pituitary gland	Ovary ● Stimulates ovulation (release of the ovum from the ovary)
Progesterone (produced if fertilised ovum implants in uterus)	Ovaries (**corpus luteum**)	Uterus ● Maintains thick uterus lining if fertilised ovum implants ● High concentrations of progesterone in pregnancy stop the cycle

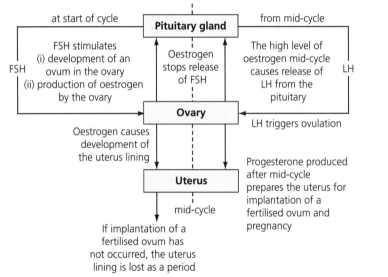

Figure 5.14 Hormones control the menstrual cycle.

> **Luteinising hormone (LH):** A hormone produced by the pituitary gland that stimulates ovulation.
>
> **Corpus luteum:** After ovulation the empty follicle turns into this and releases progesterone.
>
> **Follicle:** A structure in an ovary in which an ovum matures.

> **Revision activity**
>
> Draw out Table 5.4 with only the headings along the top and the first column on the left. Try to fill in the rest of the table from memory to help you to revise.

> **Exam tip**
>
> You should also be able to describe the roles of hormones in human reproduction including those in the menstrual cycle.

Contraception

REVISED ☐

Contraception is a name given to devices or methods that stop women becoming pregnant. It is an ethical issue because some people disagree with it for religious or moral reasons. The Catholic Church wants its followers to practice family planning or abstinence and not use other forms of contraception. Family planning involves the man not ejaculating inside the woman's vagina at the time when an ovum is likely to be in the fallopian tubes (approximately days 13 to 17 of the menstrual cycle). Abstinence is stopping having sex altogether.

A widespread form of contraception is the condom. The condom is an example of barrier contraception because it covers the erect penis and stops sperm from reaching the ovum. The diaphragm is another form of barrier contraception. It is a small plastic dome that sits at the top of the vagina covering the cervix. This stops sperm from reaching the ovum in the fallopian tubes. The contraceptive sponge is another device that sits in the same place. These devices are covered with a chemical called spermicide, which kills sperm before they can reach the ovum. **Intrauterine devices** like the coil are placed inside the uterus by a doctor or nurse and prevent an embryo from implanting.

Another method of contraception for men is having a **vasectomy**. This minor surgical operation ties knots in or cuts the sperm ducts connecting the testes with the penis. So, no sperm are ejaculated into the vagina. Tying the ducts can be undone so this process is only temporary. A similar process called **tubal ligation** in women also exists. Again tying can be undone, but cutting the fallopian tubes is a permanent form of contraception. We call this sterilisation.

Condoms have an advantage over other forms of contraception. They stop the spread of many sexually transmitted diseases (STDs) because they prevent the exchange of any body fluids.

Intrauterine devices: Contraceptive devices, such as the coil, which are placed inside the uterus to prevent an embryo from implanting.

Vasectomy: A contraceptive medical procedure during which a man's sperm ducts are blocked or cut.

Tubal ligation: A contraceptive medical procedure during which a woman's fallopian tubes are blocked or cut.

The use of contraceptive hormones

The contraceptive pill contains the hormones oestrogen and progesterone. Women take the pill at the same time each day. This may be for 21 days of the cycle, followed by a week without any pills. Other types have pills for all 28 days, but the pills taken in the final week are just sugar. This method is often easier for women because they simply take a pill every day.

The pill (as it is often referred to) stops ovulation by reducing the production of follicle-stimulating hormone (FSH). It also thickens the mucus at the cervix and reduces the thickness of the uterus lining. Both of these processes reduce the chance of pregnancy. As well as preventing pregnancy, the contraceptive pill also helps women to keep their periods more regular and can make them lighter (less blood is released).

The same hormones can be delivered into a women's blood by contraceptive patches that stick to the skin and implants that are devices which sit underneath it. Patches last for 7 days and are used weekly for the first 3 weeks. The implant can function for several years before they need to be replaced.

Exam tip

You should also be able to evaluate the different hormonal and non-hormonal methods of contraception.

⊕ The use of hormones to treat infertility

REVISED ☐

As well as preventing pregnancy, hormones can be used to help it. Some women have naturally low levels of FSH and luteinising hormone (LH) hormones. Relatively simple injections of these hormones can help some women become pregnant.

In-vitro fertilisation (IVF) is often called having a test tube baby (although no test tubes are actually used). A woman is given injections of FSH and LH. Because of this, several ova mature. These are removed in a small operation and then mixed with the father's sperm. Fertilisation therefore happens outside of the body (*in vitro*). Shortly after fertilisation, the fertilised ovum divides into an embryo by mitosis. A second minor operation places fertilised ova back into the uterus of the mother. Not all ova will embed into the lining of the uterus, so often IVF treatments involve putting back multiple fertilised ova into the uterus. The chances of having twins during IVF is reasonably high.

Fertility treatments bring huge joy to people who might not otherwise be able to have children. However, these treatments can be very emotionally and physically stressful. The success rates are not often high. The possibility of multiple births can be a risk to both the babies and mother.

Exam tip

You should be able to explain the use of hormones to treat infertility.

Negative feedback control

REVISED

The regulation of blood glucose, water content and temperature are all examples of homeostasis. Negative feedback control is a crucial part of this process. This control occurs when your body detects a change, and then makes another change to return itself to normal.

Another example of negative feedback control involves the production of the hormone thyroxine in your **thyroid gland**. Your thyroid gland determines how quickly your body uses energy, makes proteins and how sensitive it is to other hormones. So it controls your metabolic rate. Your pituitary gland produces **thyroid-stimulating hormone (TSH)**, which in turn stimulates the release of thyroxine. This is an example of negative feedback control.

Just above your kidneys are your **adrenal glands**, which produce **adrenaline**. Unlike other hormonal responses, your body responds very quickly to adrenaline. It is produced when your body perceives a threat. It is often called the 'fight or flight' response. Adrenaline increases your heart rate and provides your muscles with more oxygen and glucose for respiration. This process allows your cells to respire more, releasing more energy for 'fight or flight' (run away).

Thyroid gland: A gland in your neck that produces thyroxine to regulate how quickly your body uses energy, makes proteins and how sensitive it is to other hormones.

Thyroid-stimulating hormone (TSH): A hormone produced by your pituitary gland that regulates your thyroid gland.

Adrenal glands: Glands found in your brain that produce adrenaline.

Adrenaline: A hormone produced by your adrenal glands that causes an increase in heart rate ready for a fight or flight response.

Now test yourself

TESTED

11 In which type of structures are hormones made?
12 Name your 'master gland' and its location.
13 What happens when blood glucose is too high?
14 How are the treatments for type 1 and type 2 diabetes different?
15 State the three ways in which water is lost from the body.
H▶ 16 What hormone controls the volume of urine and where is it produced?
17 What are the names of the three processes in the kidney?
18 What advantage does a kidney transplant have over dialysis?
19 What four hormones control the menstrual cycle?
20 Which hormones are given to women before *in vitro* fertilisation?

Answers on p. 115

Exam tip

You should be able to explain the roles of thyroxine and adrenaline in the body.

Plant hormones

Control and coordination

Plants are not able to move positions like many animals, but their shoots can grow towards the light. Higher light intensities mean more photosynthesis, which means more glucose is produced. This process in turn means more respiration can occur, which results in more growth. Growing towards the light is called **positive phototropism**. Plant roots grow downwards to anchor them in the ground. This response is due to gravity and so is called **positive gravitropism** (or geotropism).

Auxins

Auxins are plant hormones that control phototropism and gravitropism. They are produced in the tips of roots and shoots. They migrate through the cells of the plant to where they are needed. Here they cause cells to grow longer in a process called **cell elongation**.

In plant shoots, auxins concentrate on the shaded side of the plant. This causes these cells to elongate, which bends the shoot towards the light. In plant roots that are growing horizontally along the ground, auxins concentrate on the sunny side of the root. This causes these cells to elongate, which bends the root downwards.

Plant roots are also able to detect and grow towards water. This process is called **hydrotropism**.

> **Positive phototropism**: The ability of plant stems to grow towards the light.
>
> **Positive gravitropism**: The ability of plant roots to grow downwards.
>
> **Auxin**: A type of plant hormone responsible for cell elongation.
>
> **Cell elongation**: The lengthening of specific cells in plants as a result of hormones.
>
> **Hydrotropism**: A plant's ability to grow roots towards water.

Experiments involving auxins

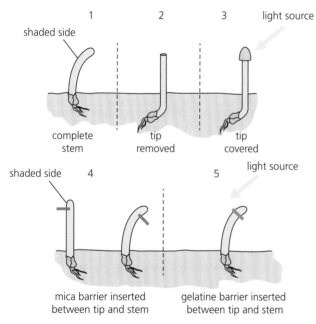

Figure 5.15 Experiments have proved that auxins are responsible for phototropism.

Figure 5.15 shows the results of investigations into phototropism. Table 5.5 (page 70) explains these results.

Table 5.5 An explanation of the results of experiments into phototropism.

Diagram	Experimental details	Results and explanation
1	No changes have been made.	This is a typical response. The shoot is growing to the light.
2	The tip of the shoot has been removed.	No phototropism occurs. This proves that auxins are produced in the tip.
3	The tip of the shoot has been covered.	No phototropism occurs. This proves that the plants' light-sensitive cells are in the tip.
4	Both sides of the tip of the shoot have had a mica (**impermeable**) barrier inserted.	No phototropism occurs when the barrier is on the shaded side. Phototropism did occur when the barrier is on the light side. This proves that auxins migrate down the shaded side.
5	The tip of the shoot has had a gelatine (permeable) barrier inserted.	Phototropism occurred. This proves that auxins can diffuse through permeable barriers like gelatine.

Other hormones

Gibberellins are a second group of plant hormones. Like auxins, they have several functions. They help with stem elongation, dormancy before seeds germinate, germination and the formation of flowers and fruit. **Ethene** is another plant hormone. It is responsible for the ripening of fruit.

Required practical 8

Seedling growth

Aim: To investigate the effect of light on the growth of newly germinated seedlings

Equipment: Radish seedlings (or similar), yoghurt pots, ruler

Method:
1 Germinate five seedlings in three different Petri dishes on damp cotton wool.
2 Draw a scientific image of the seedlings.
3 Measure and record the length of each.
4 Cover the first Petri dish with a yoghurt pot to exclude all light.
5 Cover the second dish with a yoghurt pot with a small hole in the top.
6 Cover the third with a pot with a hole in the side.
7 Record the length of each seedling again after a week.
8 Draw a second set of scientific images.

Results:
● The seedlings in the first Petri dish will have grown in length in different directions searching for the light.
● Those in the second dish will have grown directly upwards towards the light. They are likely to look healthier than the first seedlings.
● Those in the third Petri dish will have grown diagonally towards the light. They are again more likely to look healthy.

> **Revision activity**
>
> Draw out this table with only the headings along the top and the first column on the left. Try to fill in the rest of the table from memory to help you to revise.

> **Impermeable**: Not allowing substances to pass through.
>
> **Gibberellins**: Plant hormones responsible for cell elongation, seed dormancy and germination.
>
> **Ethene**: A plant hormone that ripens fruit.

Agriculture is the growth of animals or crops for food, fuels or medicines. Horticulture is the growth of plants. We use plant hormones in both industries.

Auxins are used as selective weedkillers. When sprayed on weeds in grass, more hormone lands on the larger leaves of the weeds. This causes these plants to grow uncontrollably and die.

Auxins are also used in rooting powder when plant cuttings are taken. This process is described in Figure 5.16. They encourage cells found in the stems of plants to turn into roots.

Auxins are also used in tissue culture. This is when a small number of cells are removed from a plant to grown into a genetically identical clone. They promote growth in these cells.

Ethene is used by the food industry to ripen fruit before it goes on sale. Much of our fruit is imported. It is picked before it is ripe to avoid it rotting on the journey.

Gibberellins are used to end seed dormancy, promote flowering in plants and also increase the size of fruit.

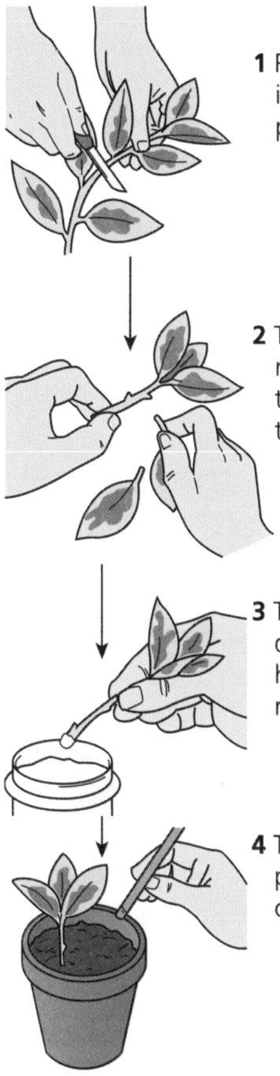

1 Part of the stem is cut from the plant.

2 The leaves are removed from the bottom of the cutting.

3 The cut end is dipped in a hormone rooting powder.

4 The cutting is planted into compost.

Figure 5.16 **How to take a plant cutting.**

Now test yourself

TESTED

21 What is positive phototropism?
22 What is positive gravitropism?
23 State two types of plant hormones.
24 How do plant shoots grow towards the light?
25 What is ethene used for?
26 Which plant hormones help with stem elongation, dormancy before seeds germinate, germination and the formation of flowers and fruit?
27 Describe the method you would use to investigate the effect of light on the growth of newly germinated seedlings.
28 State three uses of plant hormones.
29 How do auxins work as selective weedkillers?
30 How do plant hormones in rooting powder help the process of taking cuttings?

Answers on p. 115

Exam tip

You should also be able to describe the effects of plant hormones and the ways people use them to control plant growth.

Summary

- Homeostasis is the maintenance of a constant internal environment. These conditions are optimum for enzyme action and all cell functions. This includes controlling blood glucose concentration, body temperature (in warm-blooded animals) and water levels. Homeostasis can involve responses of the nervous system or hormones.
- Receptors are cells which detect changes. Coordination centres (such as the spinal cord, brain and pancreas) process information from receptors. Effectors are muscles or glands that bring about responses.
- A nerve is a bundle of neurone cells.
- The nervous system enables humans to respond to their surroundings and coordinate behaviour. Information from receptor cells passes along sensory neurones as electrical impulses to the central nervous system (CNS). The CNS is the brain and spinal cord. Within the CNS relay neurones transmit electrical impulses. Motor neurones carry electrical signals to effectors.
- Synapses are gaps between neurones which allow one neurone to pass a signal to several others. At a synapse, the electrical signal is converted into a chemical one in the form of neurotransmitters. These diffuse across the synapse and restart the electrical signal in the next neurone.
- In reflex reactions, signals are not immediately transferred to the conscious part of the brain. This saves time and potential damage to the body.
- The brain is made from billions of interconnected relay neurones. It has different regions that carry out different functions.
- It is difficult to investigate brain function and treat brain damage and disease because it is complex and delicate. Neuroscientists have mapped regions of the brain using MRI scanning techniques.
- The eye is a sense organ that contains receptors sensitive to light. The eye contains the following parts: retina, optic nerve, sclera, cornea, iris, ciliary muscles and suspensory ligaments. Accommodation is the changing of the shape of the lens to focus on near or far objects. The pupil reflex closes the size of the pupil in bright light and opens it in dim light.
- Myopia is short-sightedness. Hyperopia is long-sightedness. These conditions are treated with spectacle lenses, contact lenses, laser surgery and lens replacement.
- The endocrine system is composed of glands that release hormones into the blood. They are carried to target organs where they have an effect. These signals are slower than responses of the nervous system.
- Body temperature is monitored by the thermoregulatory centre in the brain. If this is too high, blood vessels dilate (vasodilation) and sweat is produced from sweat glands. If it is too low, blood vessels constrict (vasoconstriction) and skeletal muscles contract to make you shiver.
- The pituitary gland in the brain is the master gland. Its secretes hormones which affect other glands. Other glands include the pancreas, thyroid, adrenal gland, ovaries and testes.
- Blood glucose is monitored and controlled by the pancreas. When too high, the pancreas produces the hormone insulin which converts glucose to glycogen in the liver and muscle cells.
- When too low, the pancreas produces the hormone glucagon which converts glycogen in the liver and muscle cells to glucose. This forms a negative feedback cycle.
- Type 1 diabetes occurs when the pancreas fails to produce sufficient insulin. It is normally treated with insulin injections. Type 2 diabetes occurs when the body cells no longer respond to insulin that is produced. It is normally treated with a carbohydrate controlled diet and exercise. Obesity is a risk factor for type 2 diabetes.
- Water leaves the body from the lungs during exhalation. Water, ions and urea are all lost from the skin in sweat. Excess water, ions and urea are removed from the kidneys. The kidneys produce urine by filtration of the blood and selective reabsorption of glucose, some ions and water.
- People who suffer from kidney failure may be treated by organ transplants or using kidney dialysis.
- The digestion of proteins in diet results in excess amino acids. These are deaminated to form ammonia in the liver. Ammonia is converted to urea to be excreted.
- Anti-diuretic hormone (ADH) is produced in the pituitary gland and controls water levels. More ADH results in more selective reabsorption of water in the kidneys and so restores water levels. This is controlled by negative feedback.
- The menstrual cycle is an approximately 28-day reproductive cycle in women that runs from puberty to the menopause. Menstruation (having a period) occurs at the beginning of the cycle and last for several days. Following this, the lining of the uterus regrows in preparation for the embedding of a fertilised ovum if a woman is pregnant. Ovulation is the release of a mature ovum from an ovary which occurs around day 14. If a fertilised ovum does not embed, progesterone levels fall towards the end of the cycle and menstruation occurs.

- Follicle-stimulating hormone (FSH) causes an ovum to mature before ovulation. Luteinising hormone (LH) stimulates ovulation.
- The four hormones oestrogen, progesterone, FSH and LH interact to control the menstrual cycle.
- Contraception prevents pregnancy. Barrier contraception like condoms and diaphragms physically prevents sperm reaching the ovum. Condoms also stop the spread of STDs. Intrauterine devices prevent the implantation of the fertilised ovum. Spermicides are chemicals which kill sperm. Some people abstain from sex around ovulation. There are surgical operations for men and women which stop pregnancy.
- Hormonal contraceptives also prevent pregnancy. The contraceptive pill contains oestrogen and progesterone which inhibits FSH production which stops ova maturing. Injections, implants and patches have the same effect.
- FSH and LH can be given to women who are infertile to help them have a baby. *In vitro fertilisation* (IVF) involves the fertilisation of an ovum out of the body and its replacement into the uterus.

- Thyroxine from the thyroid gland stimulates metabolism. This is controlled by negative feedback. Adrenaline is a hormone produced by the adrenal glands. It boosts heart rate and the delivery of oxygen to the brain and muscles to prepare for a 'fight or flight' response.
- Auxins are plant hormones which are formed in the tips of shoots so they grow towards the light (phototropism) and in the tips of roots so they grow downwards because of gravity (gravitropism or geotropism). In both processes, auxins cause some cells to become longer (called cell elongation).
- Gibberellins are another plant hormone that help control seed germination. Ethene is a third hormone that ripens fruit.
- Auxins are used as weed killers, in rooting powders and for promoting growth in tissue culture. Ethene is used in the food industry to control ripening of fruit during storage and transport. Gibberellins are used to end seed dormancy, promote flowering and increase fruit size.

Exam practice

1 Which of these is NOT a use of plant hormones? [1]
   A Selective insecticides
   B Producing seedless fruit
   C Rooting powder
   D Fruit ripening
2 Where is follicle-stimulating hormone (FSH) produced? [1]
   A Ovaries   B Pituitary gland   C Uterus   D Pancreas
3 Define homeostasis. [1]
4 Describe how hormones can be used for contraception. [3]
5 Explain the process of accommodation. [4]
6 Describe what the results show for the experiments shown in the three separate diagrams. [3]

7 Describe the method used to investigate the speed of reaction times. [4]
8 Describe how your body controls its temperature. Give examples of what happens when you are too hot and too cold in your answer. [6]
9 Describe the structure and function of the nerves in the nervous system. [6]

Answers and quick quizzes online

ONLINE

6 Inheritance, variation and evolution

Reproduction

Sexual and asexual reproduction

Asexual reproduction

Asexual reproduction involves one parent organism that produces genetically identical offspring. These are called clones. There is no joining of gametes (sex cells), so there is no mix of DNA. Only mitosis is involved here. Many plants like strawberries and spider plants reproduce asexually when they produce tiny plantlets on **runners**. This is called vegetative reproduction. All bacteria reproduce asexually when they divide by **binary fission**. This process is like mitosis.

Advantages of asexual reproduction include:
- Only one parent is needed.
- It is a more time and energy efficient process as finding a mate is not required.
- It is faster than sexual reproduction.
- Many identical offspring can be produced when conditions are favourable.

Sexual reproduction

Sexual reproduction involves two parents that produce genetically different offspring. The parent organisms produce gametes which then fuse during fertilisation. Gametes are ova and sperm in animals and ova and pollen in flowering plants. So, there is a mixing of genetic information which leads to variety in offspring.

Advantages of sexual reproduction include:
- The production of **genetic variation** in offspring.
- If the environment changes, this variation gives a survival advantage by **natural selection.**
- Natural selection can be sped up by humans in **selective breeding** to increase food production.

Formation of gametes during **meiosis** requires energy. Some animals and plants produce gametes in incredibly high numbers. Many fish release millions of gametes at one time. The process of finding a mate (**courtship**) also often requires energy. Organisms that reproduce asexually do not need to use this energy.

Both sexual and asexual reproduction

Some organisms can reproduce both sexually and asexually depending upon the circumstances:
- The malarial **parasite** reproduces asexually in a human host but sexually in the mosquito.
- Many fungi can reproduce asexually by producing mitotic spores and sexually by producing meiotic spores.
- Many plants produce seeds sexually but also can reproduce asexually during vegetative reproduction.

Asexual reproduction: Reproduction involving one parent with genetically identical offspring.

Runner: An offshoot of a plant on which plantlets are produced by asexual reproduction.

Binary fission: The asexual reproduction of bacteria.

Genetic variation: Inherited differences in organisms.

Natural selection: The process by which organisms that are better adapted are more likely to survive and reproduce, which was first described by Charles Darwin.

Selective breeding: A process by which humans have chosen organisms to breed together to develop desirable characteristics, famously in dogs.

Meiosis: Cell division which forms non-identical, haploid gametes (sex cells)

Courtship: Behaviours to attract a mate.

Parasite: An organism that damages its host but depends on it to survive.

Exam tip

You should be able to explain the advantages and disadvantages of asexual and sexual reproduction.

Exam practice answers and quick quizzes at **www.hoddereducation.co.uk/myrevisionnotes**

Meiosis

Gametes are sex cells which are produced in meiosis. They are haploid and so must have half the number of chromosomes of a normal diploid body cell. A diploid human body cell like a nerve or muscle cell has 46 chromosomes, or 23 pairs. So, haploid human sperm and ova must have 23 chromosomes. During fertilisation, the sperm and ovum fuse to make a genetically different, diploid fertilised ovum. The two sets of 23 chromosomes have fused to form a new diploid cell with 23 pairs of chromosomes. This new cell divides by mitosis. Its cells can then differentiate as it grows into an adult organism.

> **Exam tip**
>
> You should be able to state that meiosis leads to non-identical cells being formed whilst mitosis leads to identical cells being formed.

The process of meiosis

The steps in the process of meiosis are shown in Figure 6.1.

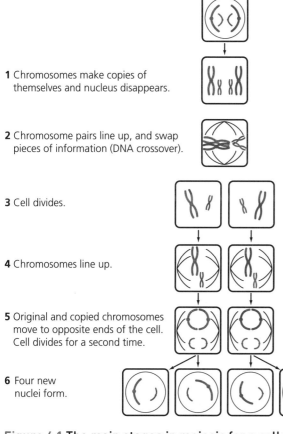

1 Chromosomes make copies of themselves and nucleus disappears.

2 Chromosome pairs line up, and swap pieces of information (DNA crossover).

3 Cell divides.

4 Chromosomes line up.

5 Original and copied chromosomes move to opposite ends of the cell. Cell divides for a second time.

6 Four new nuclei form.

Figure 6.1 The main stages in meiosis for a cell with just two pairs of chromosomes. Once the matching chromosomes have been paired up, they may swap pieces of information between them. This is called DNA crossover. The cell divides twice and ends up as four daughter cells (gametes), each with half the original number of chromosomes.

> **Exam tip**
>
> You should be able to explain how meiosis halves the number of chromosomes in gametes and fertilisation restores this to the full number.

During meiosis:
- Copies of the organism's DNA are made (a human cell would now have 92 chromosomes).
- The cell divides on two separate occasions (firstly back to two diploid cells with 46 chromosomes and then to four haploid cells with 23 chromosomes).
- DNA is exchanged to ensure all gametes are genetically different from each other.

> **Revision activity**
>
> Draw out Table 6.1 with only the headings along the top and the first column on the left. Try to fill in the rest of the table from memory to help you to revise.

Table 6.1 **The key differences between mitosis and meiosis.**

	Mitosis	Meiosis
Number of cells at beginning	One	One
Type of cell at beginning	Diploid body cell (23 pairs of chromosomes in humans)	Diploid body cell (23 pairs of chromosomes in humans)
Number of cells at end (daughter cells)	Two	Four
Type of cell at end	Diploid body cell (23 pairs of chromosomes in humans)	Haploid gamete (23 chromosomes in humans)
Number of divisions	One	Two
Identical or non-identical cells produced	Identical	Non-identical
Used for	Growth and repair	Producing gametes
Where it occurs	Everywhere except the sex organs	Sex organs (ovaries and testes in mammals)

DNA and the genome

Chromosomes and genes

Your genome is one copy of all your genetic information (your DNA). An identical copy of your genome exists in all your diploid cells because they were produced by mitosis. Your haploid gametes (sperm or ova) were produced by meiosis and so only have half of your genome.

Your genome consists of about 2 m of DNA. This fits inside the nucleus of most of your cells. (Your red blood cells do not have any DNA in them to maximise the oxygen they can carry.) To fit this length of DNA into your microscopic cells, it is arranged neatly into shapes called chromosomes. Your genome is made from 46 chromosomes which come in 23 pairs because half were present in each of the sperm and ova that made you. They paired up during fertilisation.

Chromosomes have regions that contain the DNA code to make proteins. These are called genes. Because you inherited one chromosome from each parent, you have two copies of almost all genes. These are called alleles. We think there are about 24 000 genes that carry the instructions to make our proteins. The rest of the DNA is non-coding. It does not make proteins and we call this 'junk DNA'.

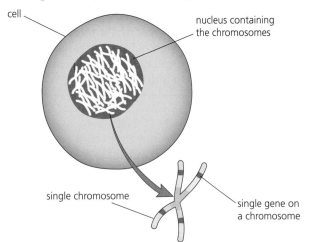

Figure 6.2 **Make sure that you can identify the nucleus, chromosomes and the gene locations on the chromosomes.**

DNA structure

DNA is a polymer made from two strands that form a **double helix**. This structure is shown in Figure 6.3. The strands are made from alternating sugar and phosphate molecules. Between these two strands are DNA bases which are joined by weak hydrogen bonds to form **complementary base pairs**. Adenine (A) and thymine (T) bases always pair together. Cytosine (C) and guanine (G) bases always pair together. Each base with its sugar and phosphate molecules is called a **nucleotide**. These are repeating units in the DNA polymer.

The sequence of bases is the code for an amino acid. The order of bases controls the order in which amino acids are assembled to make a protein.

These same four bases make up the DNA of all life on Earth. DNA is therefore universal. You have around three billion base pairs in your genome.

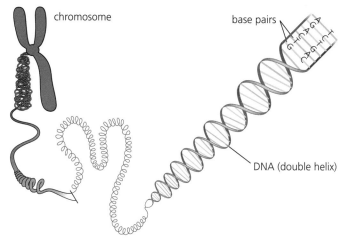

Figure 6.3 DNA is coiled into a double helix and wound into chromosomes.

Double helix: The characteristic spiral structure of DNA.

Complementary base pairs: The four letters A–T and G–C, which are always paired in DNA.

Nucleotide: A DNA base pair together with a sugar and phosphate molecule that make up the backbone of the double helix.

Cystic fibrosis (CF): A genetic disorder in which sufferers do not inherit a gene from both parents and have excess mucus in their lungs.

Exam tip

You should be able to describe the structure of DNA and define the term 'genome' as one copy of all the DNA found in your diploid body cells.

The Human Genome Project

In 2003 a nine-year project finished which involved 20 universities in countries like the USA, UK, Japan, France, Germany and China. Collectively they identified every one of the around three billion bases that made up the genome of male and female volunteers. The results of this incredible effort are stored on the internet and are available to all.

Since its completion, scientists have identified the areas of the human genome that contain genes and those areas that are non-coding. From this project we have developed tests to show the likelihood of genetic disorders like **cystic fibrosis** occurring. It is highly likely that this will have great significance for medicine in the future. It is important that we understand the human genome so we can:
- search for genes linked to different types of disease
- understand and treat inherited disorders
- trace human **migration** patterns from the past.

⊕ Protein synthesis

Proteins are made from long chains of amino acids called **polypeptides**, which are folded into the correct shape. This process ends in cell components called **ribosomes**, which are found in the cytoplasm. Proteins are important because they affect almost every

Polypeptide: A chain of amino acids that has not been folded into a protein.

part of every process within your cells. Enzymes and hormones are proteins.

Many proteins only last for several days in your body before they are broken down by your liver into urea, which is then excreted. So many of your cells are continuously undergoing protein synthesis. It is a two-step process.

Transcription

This is the first part of protein synthesis and occurs in the nucleus. An enzyme attaches to a strand of DNA just before a **gene** begins. The enzyme moves along the gene, breaking hydrogen bonds between bases, unzipping the DNA. The enzyme attracts complementary **messenger RNA (mRNA)** bases to match one side of the exposed DNA strand. These mRNA bases make an opposite copy of the DNA. Opposite a G is a C and opposite a C is a G. Opposite a T is an A, but opposite an A is a U (uracil) base. All T (thymine) DNA bases are replaced by U bases in RNA. A completed strand of mRNA might be hundreds or thousands of bases. It detaches from the DNA, which then reforms hydrogen bonds and zips up again. The mRNA strand leaves the nucleus and passes into the cytoplasm to find a ribosome.

Translation

The second part of protein synthesis begins when a strand of mRNA attaches to a ribosome. It passes through the ribosome. For every three mRNA bases, the ribosome attracts a **transfer RNA (tRNA)** molecule. The three bases on the tRNA molecule are the opposite of those of the three mRNA bases. There are only about 20 different amino acids. A chain of amino acids is called a polypeptide. The polypeptide is folded into the correct shape to make a protein.

Mutations

Changes to the sequence of DNA are called mutations. These can happen naturally, or be caused by **carcinogenic** chemicals and ionising radiation. In fact, mutations happen continuously. Mutations can involve swapping the sequence of bases, deleting them or adding them. If a mutation occurs in a gene, the sequence of DNA bases that corresponds to mRNA, then tRNA and finally the sequence of amino acids might be changed. This mutation could give the organism an advantage by making a better protein, or a disadvantage by not making a protein or making a less effective one. Here an enzyme may no longer fit the substrate binding site or a structural protein may lose its strength. Most mutations do not alter the shape of the protein and so have no effect on organisms.

Not all DNA codes for genes. These non-coding regions can switch genes on and off, so mutations in these regions may affect how genes are expressed.

Genetic inheritance

REVISED

Sexual reproduction involves two parents who produce genetically different offspring. The offspring possess genes from both parents. Some characteristics are controlled by a single gene. Examples include eye

---

Exam tip

You should also be able to discuss the importance of understanding the human genome.

Ribosome: A small organelle in the cell cytoplasm in which proteins are made.

Gene: A section of a chromosome made from DNA that possesses the code to make a protein.

Messenger RNA (mRNA) (ribonucleic acid): A chain of bases that makes an inverted copy of a gene's worth of DNA during transcription.

Transfer RNA (tRNA) (ribonucleic acid): A molecule that lines up with every three bases of mRNA to bring an amino acid to make a protein during **translation**.

Transcription: The process of making a mRNA copy of a gene sequence of DNA.

Translation: the assembly of a polypeptide chain (to make a protein) using tRNA in a ribosome

Carcinogen: A cancer-causing substance.

Revision activity

Write out the procedure for protein synthesis as a flow diagram to help you remember.

Exam tip

You should be able to explain how a change in DNA structure may result in a change in a protein. You are not expected to know the structure of mRNA or tRNA or the detailed structure of amino acids or proteins.

colour and blood group in humans. You inherit a gene for each of these characteristics from each parent. So, you have two copies of each gene. We call these alleles.

Eye colour

Each of the sperm and the ovum that made you contained a gene for eye colour. These alleles determine your eye colour. If both genes gave the same colour, say brown, it makes sense that you would have brown eyes. But what if they were different? At GCSE you only need to learn about brown and blue eyes. We use letters to represent the colours. We call these **genotypes**. **Phenotypes** are a description of a genotype using words.

Because you inherited a gene from each of your parents there are two letters in a genotype. Brown eyes are **dominant** over blue eyes. We use a capital letter to show this. Those genes that can be dominated are called **recessive**. So B is a brown eyed gene and b is a blue eyed gene. Inheriting a B from both parents (BB) is called **homozygous dominant**. Inheriting a b from both parents (bb) is called **homozygous recessive**. Inheriting one of each (Bb) is called **heterozygous**.

Table 6.2 The three possible allele combinations (genotypes) for eye colour.

Genotype	Phenotype	Terminology
BB	Brown eyes	Homozygous dominant
bb	Blue eyes	Homozygous recessive
Bb	Brown eyes	Heterozygous

We complete genetic crosses in **Punnett squares** to see the likelihood of inheriting certain characteristics. In the example in Figure 6.4 the mother's genotype is BB. This means her phenotype is brown eyes. Because she is homozygous dominant, all her ova will have B. The father's genotype is bb. Because he is homozygous recessive all his sperm will have b. So, each of the four possible combinations for their children will be Bb. That is, all their children will be heterozygous and have brown eyes.

The outcomes of these crosses can be given as percentages or ratios.

The possibilities of other characteristics such as whether you have ear lobes or can roll your tongue can be determined using Punnett squares. These are shown in Table 6.3.

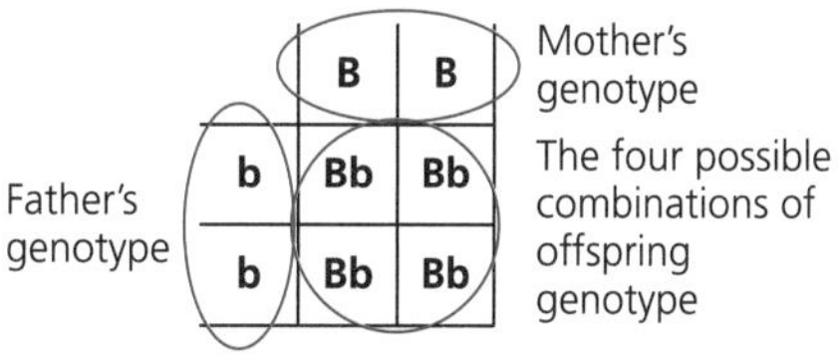

Figure 6.4 The four possible genotypic offspring of the parents whose genotypes were BB and bb.

Table 6.3 The three possible genotypes and phenotypes for ear lobes and tongue rolling.

Terminology	Ear genotype	Ear phenotype	Tongue genotype	Tongue phenotype
Homozygous dominant	EE	Free lobes	TT	Can roll
Homozygous recessive	ee	Attached lobes	tt	Can't roll
Heterozygous	Ee	Free lobes	Tt	Can roll

Exam tip

You should be able to explain all of the key terms in this section.

Genotype: The genetic make-up of an organism represented by letters.

Phenotype: The physical characteristics of an organism as described by words.

Dominant: Will show a characteristic if inherited from one or both parents.

Recessive: Will show a characteristic only if inherited from both parents.

Homozygous dominant: A genotype with two dominant alleles.

Homozygous recessive: A genotype with two recessive alleles.

Heterozygous: A genotype with one dominant and one recessive allele.

Punnett square: A grid that makes determining the chance of inheriting a characteristic easier to understand.

Revision activity

Draw out these tables with only the headings along the top and the first column on the left. Try to fill in the rest of the tables from memory to help you to revise.

Exam tip

Be careful when filling in Punnett squares that your capital and lower case letters look different. A 'C' and a 'c' (for example) can look very similar.

Most characteristics are because of multiple genes interacting, rather than a single gene as described previously.

> **Exam tip**
>
> You should be able to use direct proportions and simple ratios to express the outcomes of a genetic cross in a Punnett square and on family trees.

Family trees

A family tree can show the inheritance of characteristics over multiple generations. Every generation has its own horizontal line, with the oldest at the top.

> **Exam tip**
>
> You should be able to make **predictions** using probability for genetic crosses.

> **Prediction**: A statement suggesting what you think will happen.

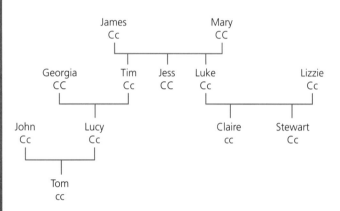

Figure 6.5 The gene for cystic fibrosis is shown by the letter 'c' in this family tree. CC is normal. Cc is a carrier who does not have the disorder but who could give it to their children. cc is a person with cystic fibrosis.

Inherited disorders

REVISED

Communicable diseases are caused by pathogens. Inherited conditions, like cystic fibrosis, are called disorders and the people who inherit them from their parents are called sufferers.

Cystic fibrosis

About one in every 10 000 people in the UK has cystic fibrosis. Sufferers inherit a recessive allele from both parents. So, all sufferers are homozygous recessive. If they had inherited one dominant gene from either parent, they would be heterozygous and not have cystic fibrosis. In this case, they would be able to pass it to their children and so we call them carriers for the disorder.

Cystic fibrosis is a disorder of cell membranes. Sufferers have excess mucus produced in their lungs, digestive and reproductive systems. This often becomes infected. Frequent physiotherapy sessions remove much of the mucus. Sadly, there is no cure for cystic fibrosis and sufferers have a much-reduced life expectancy.

Polydactyly

Polydactyly is a genetic disorder that results in an extra toe or finger. This is very rare condition that results from the inheritance of a dominant allele from one or both parents.

Sex determination

We call the 23rd pair of chromosomes the sex chromosomes because this is what they determine. Again, here we use letters. All ova are X. Approximately half of sperm are also X and the rest are Y. An X ovum and an X sperm develop into a female. An X ovum and a Y sperm develop into a male.

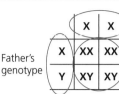

Mother's genotype

The four possible combinations of offspring genotype

Father's genotype

Figure 6.6 The relative proportions of offspring of this cross explain why approximately 50 percent of the human population is of each sex.

> **Exam tip**
>
> You should be able to carry out a genetic cross to show sex inheritance. You should be able to use direct proportions and simple ratios.

Now test yourself

1 What are the products of asexual reproduction?
2 How do strawberries and spider plants reproduce asexually?
3 Describe the advantages of asexual reproduction?
4 Define the term selective breeding.
5 Describe the advantages of sexual reproduction.
6 What are the products of meiosis in animals?
7 What are two key differences of the products of meiosis and mitosis?
8 Why is it important in medicine that the human genome is understood?
9 What happens during translation in protein synthesis?
H 10 What percentage of children will be homozygous recessive for an inherited disorder if both parents are carriers?

Answers on p. 115

Variation

Variation

Variation is the sum of all the differences between two organisms of the same or different species. So, it is all the differences between two cats (the same species) and also a cat and a dog.

Causes of variation

Variation can be caused by:
1 Environmental factors such as scars and tattoos
2 Genetic factors like blood group and eye colour
3 Both environmental and genetic factors together for your weight and height.

When genetic and environmental factors work together, a person's genome interacts with the environment to influence their development.

> **Variation:** The differences that exist within a species or between different species.
>
> **Continuous (data):** Data that come in a range and not in groups.

> **Exam tip**
>
> You should be able to describe how genetics (a person's genome) interact with the environment to influence the development of organisms.

Types of variation

The results from investigations into variation can be grouped into two types: continuous and discontinuous. **Continuous data** comes in a range and values can be halfway between them. Height is an example of this. You can be 140, 141 or 140.5 cm tall. Continuous data is presented in a line graph with a line of best fit. **Discontinuous data** comes in discrete groups. Your blood group can be A, B, AB or O. It cannot be halfway between them. Discontinuous data are presented in a bar chart.

Normally distributed variation

Line graphs of continuous data often show a characteristic 'bell-shaped graph'. This is shown in Figure 6.7. We call this a **normal distribution**. It means most values are towards the middle, and there are an ever-smaller number towards the outsides.

Mutations

A change to our DNA is called a mutation. These mutations occur naturally when our cells divide by mitosis. They occur more commonly in cells exposed to carcinogenic chemicals or ionising radiation. Very rarely this exposure can change the characteristics of organisms in a population. If these changes are advantageous, this can lead to a rapid evolutionary change.

> **Exam tip**
>
> You should be able to recall that most mutations have little or no effect on an organism's phenotype.

> **Discontinuous (data)**: Data that come in groups and not a range.
>
> **Normal distribution**: Data that are more common around a mean and form a bell-shaped graph.

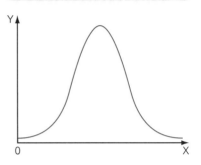

Figure 6.7 Bell-shaped graphs show normally distributed data. The most common values are in the middle and the least common values are at both ends.

Selective breeding

REVISED

All dogs are one species. That means that, regardless of their breed, they can all interbreed to produce fertile offspring. All dogs are also descended from wolves. They have not evolved from wolves but been selectively bred by humans. Before they understood the mechanisms of inheritance or the process of **evolution**, our ancestors knew that if they bred a large dog with a large bitch, they were likely to get large puppies. This selective breeding has occurred repeatedly over thousands of years and has resulted in large breeds like the Great Dane. Similarly, breeding a protective dog and a bitch over many generations has led to the development of breeds like the German Shepherd. This is **selective breeding**. It is also called **artificial selection** to remind us that it is not natural selection, which leads to evolution.

Selective breeding has given us Jersey cows, which have been selected to produce creamy milk and Friesian cows, which produce a larger volume of less creamy milk. We have also selectively bred:
- crops that are resistant to disease
- animals that produce more meat
- domestic dogs that have a gentle nature
- large and unusual flowers.

Selective breeding can lead to a reduction in the variation of a population. This is called **inbreeding** and results in genetic weakening of species. Occasionally, and by mistake, the selection of a key characteristic like size can also magnify a less desirable one. Many pedigree dogs suffer from hip misalignment because of this.

> **Exam tip**
>
> You should be able to explain the impact of selective breeding of food plants and domesticated animals.

> **Evolution**: The theory first proposed by Charles Darwin that the different species found today formed as a result of the accumulation of small advantages that were passed through generations.
>
> **Selective breeding**: Breeding animals or plants with desirable characteristics.
>
> **Artificial selection**: As in selective breeding.
>
> **Inbreeding**: Artificial selection of a small number of parents, which reduces variation.

> **Revision activity**
>
> Are you clear about the differences between artificial selection in selective breeding and natural selection in evolution?

Genetic engineering

Genetic engineering is a modern and technical process by which the genome of an organism is altered by adding a gene from another organism. This allows us to directly transfer desired characteristics into species. Organisms altered in this way are called genetically modified (GM) or **transgenic**.

This is much quicker than selective breeding. Genetic engineering is sometimes called **genetic modification**. It is an ethical issue, which means some people disagree with it for religious or moral reasons. There are many regulations around genetic engineering to tightly control it. It is illegal to genetically engineer humans, but modern medical research is exploring the possibility of using genetic engineering to overcome some inherited disorders.

Glow-in-the dark rabbits

Genetic engineering has inserted the gene that makes jellyfish glow-in-the-dark into rabbits. Enzymes were used to cut out the specific gene from the genome of the jellyfish. The genome of a rabbit embryo was then cut open using the same enzyme. Another enzyme is used to seal the glow-in-the-dark gene into the embryo. The embryo was then implanted back into the uterus of a rabbit to grow normally.

It is much harder to insert the gene into every cell of an adult organism. So, we choose to genetically engineer an embryo, which will then divide naturally into an adult organism. All cells will contain the inserted gene.

Genetically engineered crops

We have genetically engineered crops to:
- be resistant to disease
- be resistant to being eaten by insects or herbivores
- produce larger yields (e.g. bigger, better fruits).

Golden rice has been genetically engineered to contain carotene. This reduces the chance of vitamin A deficiency, which causes blindness in children. Cotton has been genetically modified to be resistant to an insect pest called the weevil. Soya has been genetically modified to be herbicide resistant.

Many people think that genetic engineering of crops has the potential to feed starving people, particularly in those countries that experience drought or **famine**. Other people think that we should not interfere with God's creatures. Others worry about the spread of genes from genetically engineering crops into wild species. We are not yet sure of the effects of GM crops on wild flowers and insects. Some people are also concerned that we have not fully explored the effects of GM crops on human health.

Genetically engineered animals

Sheep have been genetically engineered to produce proteins, such as blood clotting factors that we use in medicine, in their milk. This process is shown in Figure 6.8.

> **Genetic engineering**: A scientific technique in which a gene is moved from one species to another.
>
> **Transgenic**: Describes a genetically engineered organism.
>
> **Genetic modification**: As genetic engineering.
>
> **Famine**: An extreme shortage of food, often leading to many deaths.

Exam tip

You should be able to describe genetic engineering as a process that involves modifying the genome of an organism by introducing a gene from another organism to give a desired characteristic.

Exam tip

You should be able to describe the main steps in genetic engineering.

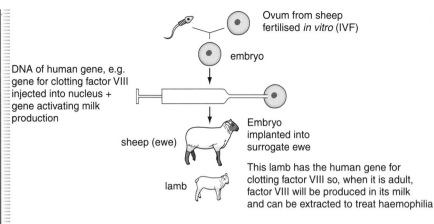

Ovum from sheep fertilised *in vitro* (IVF)

embryo

DNA of human gene, e.g. gene for clotting factor VIII injected into nucleus + gene activating milk production

sheep (ewe)

Embryo implanted into surrogate ewe

lamb

This lamb has the human gene for clotting factor VIII so, when it is adult, factor VIII will be produced in its milk and can be extracted to treat haemophilia

Figure 6.8 Genetically engineered sheep can produce human proteins in their milk.

Genetically engineered micro-organisms

We have genetically modified bacterial cells to contain the human gene for insulin. As these bacteria grow, they produce human insulin which we collect. Diabetics inject this. This process is described in detail on page 110.

Cloning

REVISED

A **clone** is a genetically identical offspring of a parent that has asexually reproduced. Many plants can reproduce asexually to make clones. Some animals, such as water fleas, possess this ability too. Clones have no genetic variation but can show differences that have resulted from **environmental variation**.

Clone: An organism produced asexually that has identical genes to its parent.

Environmental variation: Differences in organisms as a result of the environment in which they live.

Dolly the sheep

In 1996 the first mammal was cloned. She was called Dolly the sheep and was genetically identical to her one parent. This process is described in Figure 6.9.

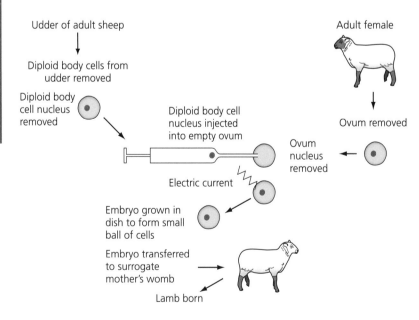

Udder of adult sheep

Diploid body cells from udder removed

Diploid body cell nucleus removed

Diploid body cell nucleus injected into empty ovum

Adult female

Ovum removed

Ovum nucleus removed

Electric current

Embryo grown in dish to form small ball of cells

Embryo transferred to surrogate mother's womb

Lamb born

Figure 6.9 The procedure for cloning sheep.

It took 277 attempts before Dolly was successfully cloned. She lived until she was 6 years old, which is less than most other sheep. Some scientists think that this was because she was cloned. Scientists have now proved that this was because of a lung infection.

Revision activity

Write out the procedure for cloning as a flow diagram to help you remember.

Other species

Before Dolly, scientists cloned tadpoles, carp fish and mice. Since Dolly we have cloned monkeys, pigs, cows and horses.

Other forms of cloning

We can also clone plants by tissue culture. During this process a small number of cells of the parent plant are removed. These are placed into a sterile growth medium and allowed to grow into a clone of the parent plant. This process is important in the preservation of rare plants. It also allows plant growers to commercially produce large numbers of genetically identical copies of prize-winning plants.

Embryo splitting occurs when a developing embryo is split apart and allowed to grow into two identical clones. **Embryo transfer** puts these embryos back into the mother's uterus to develop normally. Details of these processes are shown in Figure 6.10.

Embryo splitting: The separation of cells of an embryo to increase the number of offspring produced.

Embryo transfer: Moving fertilised embryos into other animals to increase the number of offspring produced.

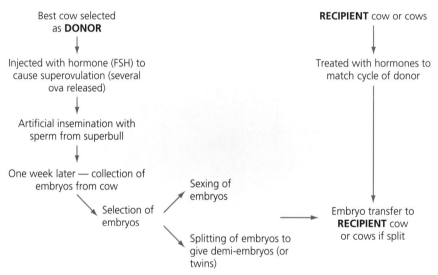

Figure 6.10 **The processes for embryo transfer and splitting in cows.**

Now test yourself

TESTED

11 Describe the causes of variation.
12 State two examples of environmental variation.
13 Define the term continuous data.
14 What type of distribution is observed in a bell-shaped graph?
15 What process have farmers used to produce Friesian and Jersey cows?
H 16 What is genetic engineering?
17 Describe why we have genetically modified crops.
18 How, specifically, has golden rice been genetically modified?
19 How was Dolly the sheep different from every other sheep before her?
20 Why do we undertake embryo splitting and transfer?

Answers on p. 115

The development of understanding of genetics and evolution

The theory of evolution

Evolution explains how the millions of different species alive today and those that have already become extinct developed from one **common ancestor**. It explains how over many generations tiny changes in individuals give them an advantage and allowed them to develop to better suit their surroundings. Eventually these differences add up to make a new species.

> **Common ancestor**: An organism from which others have evolved.

Charles Darwin and the theory of evolution by natural selection

Charles Darwin (1809–1882) is considered the father of evolution. He spent several years travelling around the world on HMS *Beagle*. During this voyage, he stopped at the Galapagos Islands off the coast of Ecuador. Here he collected animal specimens and made extensive observations. For years Darwin had developed knowledge of geology and fossils. This background, combined with his new observations, particularly of finches on the Galapagos Islands, helped Darwin develop his theory of evolution by natural selection. It says:

1 Individual organisms within a particular species show a wide range of variation within a characteristic.
2 Individuals with characteristics most suited to the environment are most likely to survive and breed.
3 These advantageous characteristics are likely to be passed to the next generation.

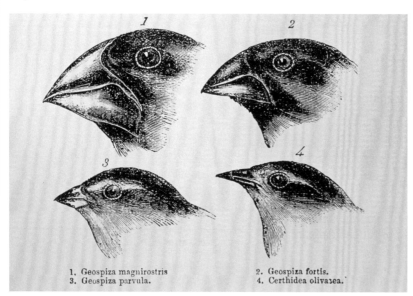

1. Geospiza magnirostris.
2. Geospiza fortis.
3. Geospiza parvula.
4. Certhidea olivasea.

Figure 6.11 Darwin's finches as drawn by Darwin himself.

Upon his return to England, Darwin was reluctant to publish his findings. He was worried about the reaction of the Church, which was very powerful at this time. He eventually published, when a colleague of his independently developed the same theory. Darwin's fears were

> **Exam tip**
>
> You should be able to describe evolution as a change in the inherited characteristics of a population over time through a process of natural selection, which may result in the formation of new species.

founded. Publication of his book *On the Origin of Species* (1859) was controversial:

- The theory challenged the idea that God made all life on Earth.
- There was insufficient evidence at the time the theory was published to convince many scientists.
- The mechanism of inheritance was only discovered 50 years after publication.

Revision activity

Are you clear about the differences between artificial selection in selective breeding and natural selection in evolution?

Lamarck and the inheritance of acquired characteristics

REVISED ☐

Jean-Baptiste Lamarck (1744–1829) developed his theory of acquired characteristics. This is often described as 'soft inheritance'. Lamarck suggested that changes that occur in an organism during its lifetime can be inherited. Scientists now know that in the vast majority of cases this type of inheritance does not occur.

Speciation

REVISED ☐

Alfred Russel Wallace (1823–1913) was inspired by Darwin's work. He too voyaged around the world. He spent eight years studying the animals and plants of what we now call Singapore, Malaysia and Indonesia. There is a narrow strait of water between these regions. Wallace noticed key differences in the species on either side of this line. It is now called the Wallace line.

From these observations, Wallace developed his own theory of evolution. Amazingly, he sent it to Darwin. At this point Darwin had not yet published his work. So, both scientists agreed to jointly published what they had discovered the following year. Despite their joint publication, Darwin is the most well remembered for evolution. Wallace is better remembered for his work on **speciation**. This is the development of different species because of evolution. He also worked on the bright warning colours of animals.

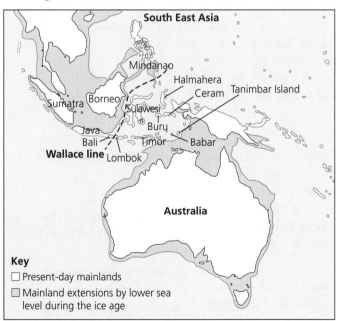

Figure 6.12 Alfred Russel Wallace discovered key differences on species on either side of the line named after him.

Speciation: The process of forming new species of life.

Exam tip

You should be able to describe the work of Darwin and Wallace in the development of the theory of evolution. You should also be able to describe the steps that give rise to new **species**.

Understanding of genetics

REVISED

Gregor Mendel (1822–1884) was a friar who spent many years studying the inheritance of characteristics in pea plants. From his studies, he developed the laws of Mendelian inheritance. These are the same rules that you use to explain how eye colour, ear lobes and tongue rolling are inherited. He saw this inheritance was due to 'units' that are passed on to descendants unchanged. We now know these units to be chromosomes, which had not yet been discovered. Mendel published his work but it was not until after his death that its significance was understood fully.

Exam tip

You should be able to describe how Mendel's work helped develop our understanding of genetics.

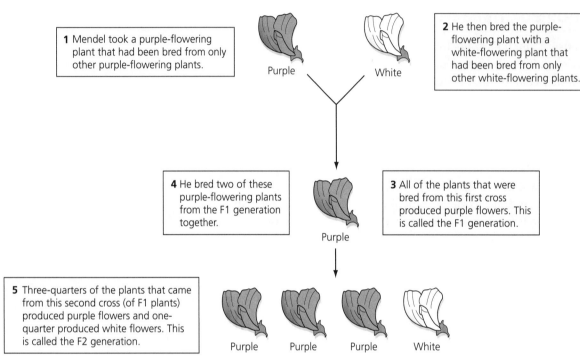

1 Mendel took a purple-flowering plant that had been bred from only other purple-flowering plants.

Purple

White

2 He then bred the purple-flowering plant with a white-flowering plant that had been bred from only other white-flowering plants.

4 He bred two of these purple-flowering plants from the F1 generation together.

Purple

3 All of the plants that were bred from this first cross produced purple flowers. This is called the F1 generation.

5 Three-quarters of the plants that came from this second cross (of F1 plants) produced purple flowers and one-quarter produced white flowers. This is called the F2 generation.

Purple   Purple   Purple   White

Figure 6.13 A flow diagram showing some of the results of Mendel's investigations into inheritance in pea plants.

Evidence for evolution

REVISED

Since Darwin published his theory of evolution, evidence for this process has developed to the point that most scientists agree with it. Most criticism of it now comes from religious groups who believe in **creationism**. This is the belief that God created the Universe and all life in it, within 7 days.

Fossils

Fossils are the remains of dead organisms preserved for millions of years in rock. Many fossils are preserved when their tissues are replaced by minerals as they decay. Other fossils are the tracks or traces of organisms. These can include dinosaur footprints, burrows and eggs. In special places like **peat** bogs, conditions are such that the rate of decay is very slow. This preserves fossils extremely well.

The **fossil record** is the information provided by all the fossils that have ever been discovered. This record can show us how significant the changes have been to species over time. There are however, gaps in the fossil record because not all fossils have been found, many have been destroyed by magma and not all parts of soft-bodied organisms become fossils.

Creationism: A belief that God created all the organisms on Earth and not evolution.

Peat: Partially decayed vegetation.

Fossil record: All of the fossils that have been discovered so far.

Antibiotic-resistant bacteria

Sir Alexander Fleming (1881–1955) discovered the first antibiotic called penicillin in 1928. Since then we have developed several other antibiotics. This process is very slow. It seems that bacteria are evolving to become immune to our antibiotics faster than we can develop new ones. A common strain of bacteria that has developed resistance is **MRSA (methicillin-resistant *Staphylococcus aureus*)**. MRSA is a communicable pathogen that kills several hundred people per year in the UK. To reduce the speed at which antibiotic resistance develops we should reduce the use of antibiotics generally and always finish the full course of medicine we are given.

Because bacteria reproduce much more quickly than many other animals, we can see the evolution of antibiotic resistance in our own lifetimes. This change provides strong evidence for evolution.

Methicillin-resistant *Staphylococcus aureus* (MRSA): A bacterium that has evolved to be resistant to antibiotics.

Exam tip

You should be able to describe the evidence for evolution including fossils and antibiotic-resistant bacteria.

Extinction

Evolution explains that organisms with characteristics most suited to the environment are more likely to survive and breed successfully. The reverse is also true. Those without these adaptations are less likely and eventually may become extinct. This occurs when there are no remaining individuals of a species still alive.

There have been several points in the Earth's history when many species have become extinct in a short period. These are called **mass extinction** events. The asteroid that killed the dinosaurs is an example of this. The latest mass extinction event is being caused by us. The rate at which extinctions are currently occurring is increasing together with an increase in the human population. We are cutting down rainforests, **overfishing** our oceans, creating bigger cities and farming more intensively. It seems that every year more species are being added to the endangered list.

Mass extinction: A large number of extinctions occurring at the same time (humans are the latest cause of a mass extinction).

Overfishing: Fishing on a scale so large that the population of species is threatened.

Exam tip

You should be able to describe factors which may contribute to the extinction of a species.

Now test yourself

TESTED

21 Define the term common ancestor.
22 Describe Darwin's theory of evolution.
23 Why did Darwin delay publication?
24 Describe the theory of evolution that Jean-Baptiste Lamarck developed.
25 Beside evolution, what else is Alfred Russel Wallace famous for?
26 Define the term species.
27 What is creationism?
28 What are fossils and how do they provide evidence for evolution?
29 What is the significance of MRSA bacteria?
30 What has been the ultimate fate of organisms that were not well adapted to their environment?

Answers on pp. 115–116

Classification of living organisms

Classification

Classification is the process by which things are placed into groups based upon their characteristics. Scientific classification of the different species of life is a very important process. Without classification, we would not be able to identify organisms. We would not know how many there were in a species. So, we would not know which ones needed conversation to protect them from becoming extinct.

Carl Linnaeus and binomial classification

Classification began with Carl Linnaeus (1707–1778). He put things into their groups based upon their structure and characteristics. He developed the **binomial** system of classification in which all living species are given a two-part name. The first is their **genus** and second their species. The binomial name for humans is *Homo sapiens*. *Homo* is our genus and *sapiens* is our species. Thousands of years ago we shared our planet with other very closely related species called *Homo neanderthal* and *Homo erectus* that are now extinct. The fact that the three species are in the same genus means they are very closely related.

A species is a group of organisms that can interbreed to produce fertile offspring. All dogs are one species so they can all interbreed. Occasionally some very closely related species can interbreed but they produce infertile **hybrids** as offspring. Horses and donkeys produce infertile animals called mules. Lions and tigers produce infertile animals called ligers.

The five-kingdom system

Linnaeus's system of binomial classification puts all life into ever bigger groups. There are five large groups called **kingdoms**. These are animals, plants, fungi, bacteria and protists. Each of these is split into smaller groups called phyla, then classes, orders and families, before genera (the plural of genus) and species.

The three-domain system

As equipment like electron microscopes and techniques involving genetic mapping of genomes has developed, so has our ability to classify organisms. Carl Woese (1928–2012) pioneered these developments and proposed the three-domain system. In this, organisms are classified into three main groups:

1 Archaea (primitive bacteria usually living in **extreme environments**)
2 Bacteria (**prokaryotes**)
3 Eukaryotes (protists, plants, fungi and animals).

Binomial: Having two names; a genus and a species (e.g. *Homo sapiens*).

Genus: The second smallest group of classifying organisms.

Hybrid: An infertile organism produced when two different species interbreed.

Kingdom: The largest group of classifying organisms, e.g. the animal kingdom.

Extreme environment: A location in which it is challenging for most organisms to live.

Prokaryotes: Prokaryotic organisms (bacteria).

Typical mistake

It is important to remember that the first part of any binomial name is the genus and the second is the species.

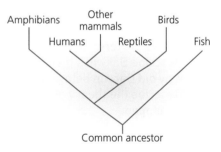

Figure 6.14 An evolutionary tree showing the relationship of some vertebrates.

Evolutionary trees

After they have been classified, the relationship between different species can be shown in diagrams called evolutionary trees.

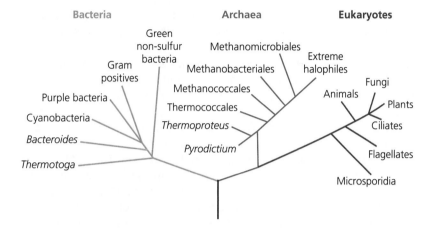

Figure 6.15 **The evolutionary tree devised by Woese and co-workers**

Now test yourself

TESTED

31 Define the term binomial classification.
32 State the binomial name for humans.
33 Which scientist developed the five-kingdom model of classification?
34 State the five kingdoms.
35 Describe the hierarchy of classification from kingdom to species.
36 What are hybrids?
37 What led to the development of the three-domain system from the five kingdoms?
38 What are the three domains?
39 In what environments are Archaea found?
40 Which scientist developed the three-domain model?

Answers on p. 116

Summary

- Like mitosis, meiosis is a type of cell division. Unlike mitosis, it produces four non-identical cells with half the DNA of the original cell. These are gametes (sex cells), which are sperm and ova in animals and pollen and ova in flowering plants.
- During meiosis, the genetic information is firstly copied. The cell then divides twice to form four non-identical daughter cells.
- Sexual reproduction involves the fusion of a male and a female gamete to restore the normal number of chromosomes. So, there is a mixing of genetic information which leads to variety in offspring. This is called fertilisation in animals and pollination in plants.
- Asexual reproduction requires one parent (not two) and so there is no fusion of gametes. Offspring are genetically identical clones of the parent. Only mitosis, not meiosis, is involved.
- Advantages of sexual reproduction include producing variation in offspring and therefore a greater survival advantage to react to changes in the environment using natural selection. Advantages of asexual reproduction include the fact that only one parent is needed, it requires less time and energy, and many identical offspring can be produced when conditions are favourable.
- DNA is a polymer made from two strands forming a double helix. In eukaryotes, this is arranged into chromosomes. A gene is a section of DNA that codes for a sequence of amino acids to make a specific protein.
- DNA is made from four complementary bases which are found in the following pairs: A–T, T–A, G–C and C–G. DNA nucleotides are made from a common sugar and phosphate group together with a base. A sequence of three bases is the code for a particular amino acid. So the order of the bases controls the order in which amino acids are assembled into polypeptides and then folded into proteins.
- Your genome is all your genetic material. The entire human genome has been sequenced in the Human Genome Project. This collaborative work was completed by 20 universities in different countries. The results are available for all on the internet. This is an important first step in the search for genes linked to different types of disease, understanding and treatment of inherited disorders, and use in tracing human migration patterns from the past.
- Protein synthesis is the manufacture of proteins which has two parts. Firstly, in transcription an mRNA copy of a gene's worth of DNA is made. The mRNA exits the nucleus.
- Secondly, in translation tRNA molecules bring amino acids to a ribosome to form a polypeptide. This is a chain of amino acids that is folded into the correct shape for a protein. In both mRNA and tRNA the base T is replaced with U.
- Enzymes, hormones and collagen are all proteins and so made in this way. Not all parts of your DNA code for proteins. Non-coding regions can switch genes on or off. Mutations in DNA can affect the order of amino acids and so the shape of a protein.
- Some characteristics are controlled by one gene such as fur colour in mice and red–green colour blindness in humans. Other examples include inheritance of eye colour, tongue rolling and the presence of ear lobes. However, most characteristics are controlled by the interaction of more than one gene.
- Alleles are pairs of genes, one inherited from each parent. The alleles present are an organism's genotype. Their expression is an organism's phenotype. A dominant allele is always expressed even if only one allele is present. A recessive allele is only expressed if two copies are present. The presence of two dominant or recessive alleles is called homozygous and the presence of one dominant and one recessive allele is called heterozygous.
- The possible outcomes of genetic crosses can be shown in Punnett squares. Inheritance of characteristics and disorders in families can be shown in family trees.
- Some medical disorders are inherited. Polydactyl is an anomaly that causes having extra fingers or toes and is caused by a dominant allele. Cystic fibrosis sufferers produce excess mucus in their respiratory, digestive and reproductive systems. This condition is caused by a recessive allele.
- Sex is determined by the combination of the 23 pairs of chromosomes. All ova are X. Half of sperm are X and half are Y. XX is female and XY is male.
- Variation is the sum of all the differences between two organisms of the same or different species. The causes of variation are genetic (e.g. eye colour and blood group), environmental (e.g. scars and tattoos) and combinations of both (e.g. height and weight).
- Data resulting from surveys into variation come in a range (continuous) or groups (discontinuous). Many biological features are normally distributed and give characteristic bell-shaped graphs.

- There is usually extensive variation within a species. This arises from mutations. Most of these have no effect on the phenotype, some influence and very few determine phenotype. Advantageous mutations can lead to relatively quick changes in species.
- Evolution is a change in the inherited characteristics of a population over time through the process of natural selection which may result in the formation of a new species (speciation). This occurs when the phenotypes of two populations mean they can no longer interbreed.
- Darwin's theory of evolution by natural selection states that all species of life evolved from simple organisms.
- Selective breeding occurs when humans select individual plants and animals to breed for their particular genetic characteristics. This is repeated over many generations. Examples include disease resistance in food crops, animals which produce more meat or milk, domestic dogs with a gentle nature and large or unusual flowers. Selective breeding can lead to inbreeding, which can magnify the incidence of disease or inherited defects.
- Genetic engineering is the process of modifying the genome of an organism by the introduction of a gene from another organism to give a desired characteristic. Plant crops have been genetically engineered to be resistant to disease and increase yield. Bacterial cells have been genetically engineered to produce human insulin to treat diabetes. Genetic engineering is an ethical issue.
- **H** In genetic engineering, enzymes are used to isolate and cut out the required gene. This gene is then inserted into a bacterial plasmid or virus. This vector is then used to insert the gene into the cells of animals, plants or micro-organisms so they develop desired characteristics.
- Tissue culture, taking cuttings and embryo transplants are all examples of cloning. Dolly the sheep was the first mammal cloned using the process of adult cell cloning. In this, the nucleus is removed from an unfertilised ovum. The nucleus from an adult body cell, such as a skin cell, is inserted into the empty ovum. An electric shock stimulates the fertilised ovum to divide by mitosis to form an embryo. This is inserted into the uterus of an adult female to develop.
- Charles Darwin developed his theory of evolution by natural selection during an around-the-world expedition. He delayed publication of his theory because of his concern about the reaction of the Church.

- Darwin's theory says individual organisms within a species show a wide range of characteristics. Individuals with characteristics most suited to their environment are more likely to survive and breed. Their offspring are likely to inherit these characteristics.
- Jean-Baptiste Lamarck developed his theory of acquired characteristics (also called soft inheritance) which suggests that changes that occur in an organism during its lifetime can be inherited. We now know that this is infrequent.
- Alfred Russel Wallace independently proposed and jointly published the theory of evolution by natural selection with Darwin. Wallace also published work that described warning colouration in animals and his theory of speciation.
- In the 19th century, Gregor Mendel carried out breeding experiments in plants. He determined 'units' of inheritance that we now know to be chromosomes. The importance of his work was not discovered until after his death.
- Fossils are the 'remains' of organisms from millions of years ago. The fossil record provides evidence for evolution. There are gaps in the fossil record because many early forms of life were soft-bodied, not all fossils have been found and many have been destroyed by geological processes. The development of antibiotic-resistant bacteria like MRSA is also evidence for evolution.
- Extinction occurs when there are no remaining individuals of a species still alive.
- To reduce the speed at which bacteria are becoming antibiotic resistant, doctors should not prescribe antibiotics unnecessarily, patients should complete their full course and the use of antibiotics in agriculture should be restricted.
- Classification is the process of grouping organisms. It is an essential step before conservation of endangered organisms can occur.
- Classification began with Carl Linnaeus in the 18th century. He classified species into a kingdom, phylum, class, order, family and genus. He developed the binomial system of naming organisms. The first part of the name is the genus and the second part is the species.
- As a result of advances in microscope and DNA sequencing technology, Carl Woese proposed his 'three-domain' system. Organisms are either *archaea* (primitive bacteria), bacteria or eukaryotes.
- Evolutionary trees show the relationship between the evolution of organisms.

Exam practice

1 What cause of variation are scars and tattoos examples of?　　　　　　　　[1]
   (a) Continuous
   (b) Genetic
   (c) Discontinuous
   (d) Environmental

2 What does the second part of a binomial name indicate?　　　　　　　　　[1]
   (a) Kingdom
   (b) Class
   (c) Genus
   (d) Species

3 Define the term speciation.　　　　　　　　　　　　　　　　　　　　　　[1]

4 Complete this Punnett Square for eye colour.　　　　　　　　　　　　　　[2]

		Female alleles	
		B	b
Male alleles	B		
	b		

5 Describe the differences in the cells produced in mitosis and meiosis.　　　　[4]

6 What does this image tell you about the evolutionary relationships between the six classes of vertebrate?　　　　　　　　　　　　　　　　　　　　　　　　　　[4]

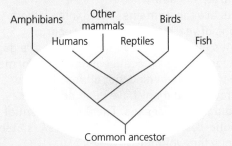

Amphibians　Other mammals　Birds
Humans　Reptiles　Fish
Common ancestor

7 Draw a graph of the results from a survey into blood groups shown in the table below.　[6]

Blood group	A	B	AB	O
Number of people	21	5	2	24

8 A man heterozygous for eye colour marries a homozygous recessive woman. Complete a Punnett square to show their possible offspring. Analyse your results as percentages and proportions.　[6]

		Female alleles	
Male alleles			

9 Doctors are now prescribing fewer antibiotics to reduce the evolution of antibiotic-resistant bacteria. Describe the process of evolution of antibiotic bacteria.　[6]

Answers and quick quizzes online

ONLINE

7 Ecology

Adaptations, interdependence and competition

Communities

A **population** is the total number of all the organisms of the same species in a geographical area. This can either be the whole planet or a much smaller part of it. A **community** is a group of two or more populations of different species in the same geographical area.

An **ecosystem** is the interaction between a community of living organisms and the non-living parts of their environment. To survive and reproduce, all organisms need resources from their surroundings and the other organisms that live with them.

In all communities, populations of organisms compete for resources. If this competition is within one population of organisms it is called intraspecific **competition**. If it is between two populations, we call it interspecific competition. Plants compete for light, space, water and nutrients from the soil. Animals compete for food, mates and territory.

> **Population**: The total number of all the organisms of the same species or the same group of species that live in a particular geographical area.
>
> **Community**: A group of two or more populations of different species that live at the same time in the same geographical area.
>
> **Ecosystem**: A community of living organisms in their environment.
>
> **Competition**: The contest between organisms for resources such as food and shelter.

Interdependence

All organisms within a community depend upon each other. This can be for food, shelter, pollination and seed dispersal. They have evolved to do this. We call this **interdependence**. A stable community is one in which high levels of interdependence are found. Here we see balance between predators and prey. The numbers rise and fall, as shown in Figure 7.1. But interdependence means they never get to the point at which one organism kills or totally outcompetes another. Removal of one species from an ecosystem can affect all other organisms within it.

> **Interdependence**: All the organisms in a community depend upon each other and because of this changes to them or their environment can cause unforeseen damage.

Exam tip

You should be able to describe the importance of interdependence and competition in a community.

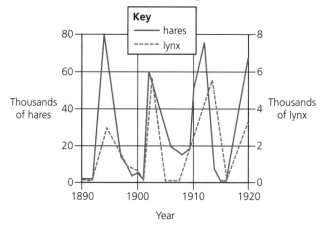

Figure 7.1 Predator–prey cycling in the Canadian lynx and snowshoe hare populations.

Abiotic factors

Abiotic factors are non-living. They can be chemical or physical, but not biological. Important abiotic factors for plants and animals are: light intensity for photosynthesis, temperature, moisture levels, soil pH and mineral content, wind intensity and direction and carbon dioxide for plants and oxygen levels for aquatic animals.

Abiotic factors: The non-living parts of the environment.

> **Exam tip**
>
> You should be able to explain how a change in an abiotic factor could affect a community.

Biotic factors

Biotic factors are living. Important biotic factors include the availability of food, numbers of new predators, introduction of a pathogen that causes a communicable disease and competition between species. Biotic factors often result from newly introduced species. If these factors damage the local ecosystem by outcompeting existing species, they are called **invasive species**. The cane toad in Australia and grey squirrel in the UK are invasive species.

Biotic factors: The living parts of the environment.

Invasive species: An organism that is not native and causes negative effects.

> **Exam tip**
>
> You should be able to explain how a change in a biotic factor could affect a community.

Adaptations

All organisms have adaptations that enable them to survive in the conditions in which they normally live. They allow individuals to outcompete others and provide them with an evolutionary advantage. Without this and the resulting competition, there would be no evolution. Adaptations can be structural, behavioural or physiological.

Structural adaptations are physical features that allow competition. They include good eyesight, sharp teeth or claws and shells or exoskeletons for protection. **Behavioural adaptations** are specific behaviours that give an organism an advantage. Animals like crows, primates, elephants and alligators use tools. This is an obvious behavioural advantage. **Physiological adaptations** are processes that help organisms. The formation of poisons by some snakes, spiders and plants is an example.

Structural adaptation: An advantage to an organism as a result of the way it is formed, like the streamlining seen in fish.

Behavioural adaptation: An advantage to an organism as a result of behaviour, such as a courtship display.

Physiological adaptation: An advantage to an organism as a result of a process, such as the production of poisonous venom.

> **Exam tip**
>
> You should be able to suggest how organisms are adapted to the conditions in which they live.

> **Typical mistake**
>
> It is important that you can explain why an organism is adapted not just state its adaptation.

Extreme environments

Extreme environments have conditions in which most life finds it hard to survive. They have:
- extremes of pH or temperature
- low levels of oxygen or water
- high salt concentration.

Deserts are extreme environments because they are very hot during the day and cold at night. They are also extremely dry. The polar regions are also extreme environments because of their cold temperature, low light level in winter and the lack of fresh, liquid water.

Deep-sea **hydrothermal vents** are gaps in the seabed where magma provides heat. Surrounding them are communities of life which exist nowhere else on Earth. These are the only food chains that do not start with a photosynthesising plant or alga. Bacteria feed on the chemicals from the vents and support all other species of life. Organisms that live in extreme environments are called **extremophiles**. Near the vents, the temperatures are very high but further away they very quickly fall. Hydrothermal vents are always dark and under high pressure.

> **Hydrothermal vents**: Volcanic vents at the bottoms of seas and oceans, where unique species of life have evolved based upon bacteria feeding on chemicals and not photosynthesis.
>
> **Extremophile**: An organism that lives in an extreme environment.

Now test yourself

TESTED

1 Define the term population.
2 Define the term community.
3 What do animals compete for?
4 What is interdependence?
5 State two examples of abiotic factors that could affect the distribution of plants in the environment.
6 What biotic factors could affect the distribution of animals in the environment.
7 State an example of a physiological adaptation.
8 What conditions exist in extreme environments?
9 What term do we give to organisms that live in extreme environments?
10 What is unusual about food chains that surround hydrothermal vents?

Answers on p. 116

Organisation of an ecosystem

A population is the total number of all the organisms of the same species in a geographical area. A community is a group of two or more populations of different species in the same geographical area. An ecosystem is the interaction between a community of living organisms and the non-living parts of their environment. A habitat is the natural environment in which an organism lives.

Sampling

REVISED

Sampling is the process of looking at a small part of an ecosystem and drawing conclusions about the whole. This saves important time and money.

> **Sampling**: The process of recording a smaller amount of information to make wider conclusions.
>
> **Quadrat**: A square frame used in biological sampling.

Quadrats

Quadrats are squares of wire that are often 0.25 m². They are often used during sampling to record the number of organisms in a specific area (within the quadrat). Quadrats are used in three ways:

● The number of a single species within them is counted.
● The number of different species are counted (a measure of biodiversity).
● The percentage cover of one species such as grass is recorded.

There are two main ways in which you use quadrats depending upon what it is you are trying to investigate. If you want to know the numbers of a species in an area, or to compare two or more areas you would place your quadrats randomly. If you want to investigate the change in a habitat you would place your quadrats systematically along a line called a **transect**.

In all cases of sampling, it is important you record more than three concordant results. It is likely you will place at least 20 quadrats and calculate mean values from your results.

Random sampling using quadrats

You cannot stand in the middle of the habitat you wish to sample and throw your quadrat. It is important that the placement is totally random to avoid bias in your results. The method you use is described below in Required practical 9 'Sampling'.

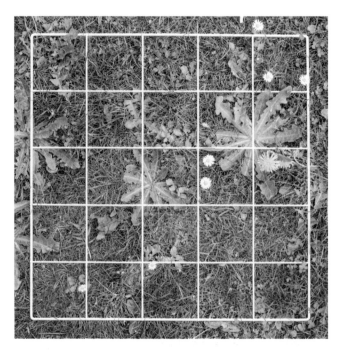

Figure 7.2 **A quadrat on a lawn with weeds to count.**

Required practical 9

Sampling

Aim: To measure the population size of a common species in a habitat. Use sampling techniques to investigate the effect of a factor on the distribution of this species.

Equipment: Quadrat, ruler

Method:
1 Choose a starting location in one corner of your ecosystem.
2 Generate random numbers using a table or calculator.
3 Use two random numbers as coordinates to tell you where to place your first quadrat.
4 Record the total number of individuals of the species you are investigating in this quadrat.
5 Return to your original starting position and repeat using different random numbers for coordinates.

Results:
Calculate a mean value for your results.

If you have counted the number of a species within your quadrat, you can estimate how many are in your habitat. To do this:
● Measure the area of your habitat.
● Multiply the mean number of organisms per quadrat by the difference in size between your quadrat and habitat.

> **Transect:** A line along which systematic sampling occurs.
>
> **Systematic sampling:** The regular distribution (not random) of a survey to answer a specific question, usually about a trend.

Systematic sampling using quadrats

Systematic sampling looks for changes in the distribution of organisms as a result of changes within a habitat. Because of this, other abiotic factors like light intensity or moisture levels are often recorded alongside biotic ones.

Exam practice answers and quick quizzes at **www.hoddereducation.co.uk/myrevisionnotes**

If you wanted to see if the number and species of seaweed changed as you walked down a seashore you would use systematic sampling. (This is different from the total number of seaweed on the shore, for which you could use random sampling.) Draw an imaginary line called a transect down your habitat. Place your quadrat at equal and regular distances down the transect. Record the number and species of seaweed in each quadrat. At the same time, record abiotic factors that might help explain the changes you observe. (In the case of the seashore, the position of seaweed is determined largely by the number of hours any species can be out of water. Those at the top of the shore have evolved to be out of the water for longer.)

Exam tip

You should be able to explain the principles of sampling.

Producers, consumers and decomposers

REVISED

All organisms need energy to complete the seven life processes: movement, reproduction, sensitivity, nutrition, excretion, respiration and growth.

Producers

Producers are photosynthetic plants and algae found at the lowest **trophic level** of almost all food chains. During photosynthesis, these species turn carbon dioxide and water to glucose and oxygen using sunlight. This glucose then supports all life at higher trophic levels. Unusually, bacteria in deep-sea hydrothermal vents are also producers. They feed on the chemicals released from the volcanic vents, which are in turn consumed by all other species in these very specific food chains. Producers are called **autotrophs** because they 'feed' themselves.

Consumers

Consumers are organisms that obtain their energy by eating others. All animals are consumers. Any that eat photosynthetic plants or algae are called primary consumers. They are also called herbivores. Consumers that eat primary consumers are called secondary consumers. Then comes tertiary and quaternary consumers as we move up the trophic levels. At the top of all food chains is the **apex predator**. Consumers that kill and eat other animals are predators and those that are eaten are prey. Consumers obtain their food from other organisms and are called **heterotrophs**.

Interestingly, at each trophic level only 10% of the energy from the previous one is passed along. So, when cows eat grass, or lions eat zebras they only obtain 10% of the total energy. The rest is used by the organisms (grass or zebras) to complete the seven life processes.

Producer: Any organism that photosynthesises (a plant or alga).

Trophic level: A stage in a feeding relationship representing an organism in a food chain or a group of organisms in a food web.

Autotroph: An organism that makes its food from simple organic compounds in its surroundings, often using energy from light.

Consumer: Any organism in a feeding relationship that eats other organisms for food.

Apex predator: The final organism in a feeding relationship.

Heterotroph: An organism that obtains its food from other organisms, e.g. humans.

Exam tip

You should be able to recall that photosynthetic organisms are the producers of most biomass for life on Earth.

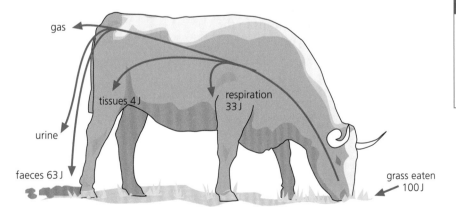

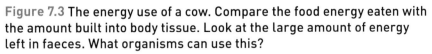

Figure 7.3 The energy use of a cow. Compare the food energy eaten with the amount built into body tissue. Look at the large amount of energy left in faeces. What organisms can use this?

Decomposers

Decomposers break down the remains of dead organisms. This process is called decomposition or rotting. Bacteria and fungi are important decomposers. Decay is slowed by high or low temperatures and in the absence of water and oxygen. Decomposing bacteria and fungi find it harder to live in these conditions.

Gardeners and farmers try to provide the optimum conditions for rapid decomposition of waste material like animal dung. This can be used to produce compost which is a natural fertiliser for garden plants and crops.

If decay occurs in anaerobic conditions, it produces methane gas. This occurs in biogas generators which produce methane as a fuel.

> **Exam tip**
>
> You should be able to explain how temperature, water and available oxygen affect the rate of decay. You should also be able to calculate the rate of decay.

Required practical 10

Decay

Aim: To investigate the effect of temperature on the rate of decay of fresh milk by measuring pH change

Equipment: Fresh milk, waterbaths set to 10, 20, 30 and 40 °C, four beakers, universal indicator paper, clingfilm

Method:
1 Place 25 cm³ of milk into each beaker.
2 Cover with clingfilm and place all beakers into a different water bath.
3 Test the pH using universal indicator every day for 5 days.

Results: As milk decays, its pH becomes more acidic. Lipids break down into fatty acids and glycerol. The bacteria in the milk also breaks down lactose sugar into lactic acid.

Materials recycling   REVISED

All materials in the living world are recycled to provide the building blocks for future organisms. That is, some of your atoms were incorporated into previous living organisms. They will also be incorporated into other organisms after you die.

The carbon cycle

The carbon cycle shows the various carbon compounds that it can form and how it is converted between them. It returns carbon from organisms to the atmosphere as carbon dioxide to be used by plants in photosynthesis.

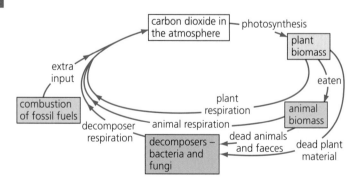

Figure 7.4 **The carbon cycle.**

The three key processes of the carbon cycle are shown in Table 7.1.

Table 7.1 **The key processes and conversions of the carbon cycle.**

Process	Carbon starts as	Carbon ends as
Photosynthesis	Carbon dioxide	Glucose
Respiration	Glucose	Carbon dioxide
Combustion (burning)	Fuel	Carbon dioxide

Revision activity

Draw out this table with only the headings along the top and the first column on the left. Try to fill in the rest of the table from memory to help you to revise.

The water cycle

The water cycle shows all the pathways water takes as it cycles through our atmosphere, rivers, lakes, seas and oceans. It provides fresh water for plants and animals on land before draining into the seas.

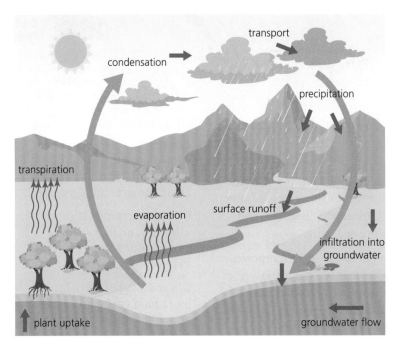

Figure 7.5 **The water cycle.**

Precipitation is the scientific name for rain, snow, hail and sleet. **Runoff** occurs when water moves across the surface of land. This is often in streams and rivers. **Infiltration** is the movement of water downwards from the surface to become groundwater. Evaporation is the conversion of liquid water to gas. Water evaporates from all surface water. Transpiration is the movement of water though plants from roots to its leaves. Water vapour evaporates from tiny pores in leaves that are called stomata.

Combustion: Burning.

Precipitation: Rain, snow, hail and sleet.

Runoff: The movement of water across the surface of land.

Infiltration: The movement of water into the ground to become groundwater.

Ⓗ Impact of environmental change of distribution of organisms

REVISED

The distribution of organisms in an ecosystem is dependent upon many, often interacting factors.

Water

Water is an essential component of all life. Without it life cannot exist. So, the abundance of water plays a crucial role in the distribution of organisms. Generally, more biodiverse ecosystems have more available water. The rainforest is more biodiverse than the desert.

Some organisms that live in freshwater can tell us how polluted it is. These are called **bioindicators**. Their presence or absences tells us about pollution levels. They are shown in Figure 7.6.

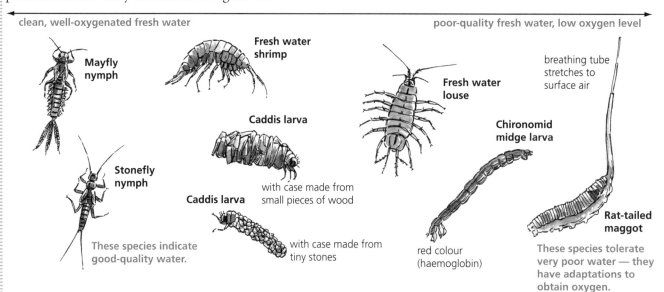

clean, well-oxygenated fresh water → poor-quality fresh water, low oxygen level

Mayfly nymph

Fresh water shrimp

Fresh water louse

breathing tube stretches to surface air

Caddis larva

Chironomid midge larva

Stonefly nymph

Caddis larva

with case made from small pieces of wood

These species indicate good-quality water.

with case made from tiny stones

red colour (haemoglobin)

Rat-tailed maggot

These species tolerate very poor water — they have adaptations to obtain oxygen.

Figure 7.6 Species that indicate the quality of water (bioindicators).

Temperature

Organisms have evolved to function best at specific temperatures. Human body temperature is 37 °C. Deviation from this means that our enzymes are less efficient, so reactions happen slower. Ultimately, increases and decreases in temperature can be fatal if not addressed.

Global warming is the gradual increase of the Earth's average temperature. This warming is caused by the **greenhouse effect**. If temperatures continue to increase, organisms that could survive in the tropics will eventually migrate towards the poles to find these temperatures. How long will it be before mosquitos carrying malaria live in the UK?

Atmospheric gases

The combustion of fossil fuels releases carbon dioxide, and oxides of sulfur and nitrogen into the atmosphere. **Lichens** are excellent bioindicators of air pollution. These organisms are a combination of a fungus and an alga which live together in a mutualistic relationship.

> **Bioindicator**: An organism whose presence or absence tells you about the cleanliness of an ecosystem.
>
> **Global warming**: The effects of an increase in temperature as a result of the greenhouse effect.
>
> **Greenhouse effect**: Increased carbon dioxide and other gases trap more of the Sun's solar radiation in the Earth's atmosphere, leading to global warming.
>
> **Lichen**: A symbiotic growth of a fungus and an alga, which are bioindicators for clean air.

Table 7.2 Common lichens that act as bioindicators of air pollution from sulfur dioxide.

Type of lichen	(1) Common orange lichen, quite tolerant of moderate pollution levels	(2) Quite a common lichen, but dies quickly if pollution levels rise	(3) A common woodland lichen and an indicator of clean air	(4) Beard lichen, only survives in pure air

TESTED

Now test yourself

11 Define the term sampling.
12 State the three ways in which quadrats are used.
13 When would a transect be used instead of random sampling?
14 What term describes an organism that makes its food from simple organic compounds in its surroundings, often using energy from light.
15 What proportion of energy is transferred between tropic levels?
16 Under what conditions does decomposition slow?
17 State the three main processes of the carbon cycle.
18 What is precipitation?
19 What are bioindicators? Give an example of a bioindicator of air pollution in your answer.
20 Define the term global warming.

Answers on p. 116

> **Exam tip**
>
> You should be able to evaluate the impact of environmental changes on the distribution of species in an ecosystem.

Biodiversity and the effect of human interaction on ecosystems

Biodiversity

REVISED

Biodiversity is a measure of the variety of all the different species of organisms on Earth, or within a particular ecosystem. Areas of low biodiversity include deserts and the polar regions. Areas of high biodiversity include tropical rainforests and ancient oak woodlands.

A relatively high biodiversity for any ecosystem ensures its stability by reducing the dependence of one species on another for food, shelter and the maintenance of the physical environment. Crucially, the future of the human species on Earth probably depends upon us maintaining a good level of biodiversity. Many of our activities, including **deforestation**, pollution and burning of fossil fuels, reduce biodiversity. Are we doing enough?

> **Deforestation**: Cutting down of trees (often on a large scale).

Waste management

REVISED

The human population has increased past seven billion. One hundred years ago, there was less than two billion people. This rapid growth in our population and an increase in our standard of living has meant that we are using evermore resources. We are also producing evermore waste. Unless our waste is properly disposed of, evermore pollution will be produced.

Water pollution

All life needs water. We used water for drinking, growing our food, washing, transportation and to find food. Pollution of water hurts us and the animals and plants that live in or near this water that we also depend upon. It is estimated that over one billion people do not currently have access to clean water.

Water pollution comes from pathogens like *Salmonella* bacteria, *Norovirus* and parasitic worms. We often find these in water polluted by sewage. Fertilisers can also wash from farmer's fields in heavy rains and pollute nearby streams and rivers. This can lead to **eutrophication**. Some

> **Eutrophication**: Death of all life in an aquatic ecosystem as a result of overuse of fertilisers often on nearby farmland.

factories still illegally release toxic chemicals into rivers and oceans. Some of these, like the **pesticide** DDT and metal mercury, cannot be easily excreted. So they concentrate at higher trophic levels. We call this **bioaccumulation**. Oil spills have drastically polluted some coastal areas killing hundreds of thousands of sea birds, otters and seals.

Air pollution

Air pollution is often caused by waste gases from vehicles or factories. Without suitable regulations, excessive particulates produced from these sources can cause **smogs** that are seen covering large cities in some parts of the world. Sulfurous oxides are produced by the burning of fossil fuels. They can react with water vapour to form **acid rain**. They can destroy entire forests of trees and damage stone buildings and statues. Carbon monoxide is another polluting gas. This poisonous gas is produced during **incomplete combustion**. It is odourless and replaces oxygen in your red blood cells. It slowly suffocates you.

Land pollution

In recent years, our society has become more consumable. We seem to be mending fewer things and just buying more to replace them. This change has resulted in a significant increase in the rubbish we produce. Despite the efforts of local councils, a significant amount of this rubbish is still not recycled and goes into landfill. As our waste slowly rots away it can produce a toxic liquid called leachate and release large volumes of greenhouse gases. To reduce landfill, it is essential that we reduce our consumption, reuse as much as we can before buying new and **recycle**. (These are called the three R's.)

Pesticides: Chemicals used to kill pests.

Bioaccumulation: The increase in concentration of toxins at higher trophic levels in a food chain.

Smog: Fog or haze because of smoke or other polluting gases.

Acid rain: Precipitation that is acidic because of air pollution.

Incomplete combustion: The burning of fuel without sufficient oxygen, which produces poisonous carbon monoxide.

Recycle: Changing a waste product into a new raw material to make another.

Land use
REVISED ☐

The way we use our land is called land use. This can be for farming crops (**arable**) or rearing animals. It can be forests of trees to produce timber. It can be urban for towns and cities. We quarry or mine some areas and dump waste in others. It can also be preserved in national parks or other conservation areas. During the Stone Age, much of the UK was covered in forests and very little was farmland. Think about the UK now. The clear majority of it is now farmland or urban towns and cities. Our land use has changed dramatically in this time.

Arable: Farming of crops for food.

Biodiversity: A measure of the different species present in a community.

Sink: A long-term store of a substance, often carbon.

Peat bogs

Peat bogs have wet, acidic soils with low levels of nutrients. Very few trees grow in them. They have relatively low **biodiversity** but contain species that are not found in other places. Because of these conditions, very little decomposition occurs in them and peat accumulates. It is partially decayed vegetation. Peat can be the first step in the slow process of forming fossil fuels. In some parts of the world, peat is dug up and used as a fuel. This releases lots of carbon dioxide into the atmosphere that was stored as a carbon 'sink' in the peat. The destruction of peat bogs reduces biodiversity.

Deforestation
REVISED ☐

Deforestation is the cutting down of trees so that an area can be used for other purposes, often to make farmland for cattle or rice fields, or grow biofuels. Deforestation began around 12 000 years ago as we turned from **hunter-gatherers** to farmers. However, in recent years the rate at which deforestation has occurred has increased massively. Now huge areas of rainforest are cut down to grow crops like palm oil. It therefore drastically reduces biodiversity. We have cut down over half of the rainforest that existed 75 years ago.

Deforestation stops plants photosynthesising and so they do not remove carbon dioxide from the atmosphere. The wood from deforestation is often burned which releases more carbon dioxide into the atmosphere. Deforestation therefore increases the greenhouse effect and global warming.

Global warming

Global warming is the gradual increase in the Earth's average temperature. This has changed naturally over time. We have had ice ages and other times when tropical conditions were present in the UK. However, almost all scientists agree that the current rate of change is faster than the planet has ever seen before. They also almost all agree that this is occurring because of the greenhouse effect.

The effects of global warming include the melting of glaciers and polar icecaps, raising ocean levels threatening low lying cities like London and New York, freak weather and species migration. How long will it be before global warming threatens **food security** in the UK?

Greenhouse effect

Greenhouse gases include carbon dioxide and methane. As we release larger volumes of greenhouse gases, they are trapping more heat in our atmosphere. At low levels, the greenhouse effect is needed to keep our planet warm enough to support life. However, analysis of drilled ice cores from the polar regions has shown our human activity in the last few hundred years has doubled the carbon dioxide in our atmosphere. It is essential that our governments, companies and voluntary organisations, and we as individuals, all act to stop this increase before it is too late.

Maintaining biodiversity

Biodiversity is a measure of the numbers of different species on our planet or in an ecosystem. **Conservation** is one way in which biodiversity is maintained. This protects certain areas with high ecological importance like peat bogs, ancient forests and marine regions like coral reefs. Other ways of maintaining biodiversity include:

- zoo **breeding programmes** to increase the number of endangered species
- reintroduction of hedgerows in areas that have been **intensively farmed**
- reducing deforestation and carbon dioxide emissions by some governments
- reducing, reusing and recycling rather than dumping in landfill.

21 What is biodiversity?
22 Why do some pesticides bioaccumulate?
23 What process produces poisonous carbon monoxide?
24 What three 'R's' help live a more sustainable lifestyle?
25 What conditions are found in peat bogs?
26 What are carbon 'sinks'?
27 Describe the effects of global warming.
28 How is biodiversity maintained?
29 Define the term food security.
30 What is intensive farming?

Answers on pp. 116–117

> **Exam tip**
>
> You should be able to describe both positive and negative human interactions in an ecosystem and their impacts on biodiversity.

Trophic levels in an ecosystem

Trophic levels   REVISED ☐

Food chains show the feeding relationships between organisms in an ecosystem.

grass → cow → human

So, in the food chain above, humans eat cows which eat grass. The arrows show the flow of energy from one organism to another. A trophic level is any level in a food chain. Levels in a food chain almost always follow this pattern:

- The first trophic level is plants and algae which make their own food by photosynthesis.
- The second trophic level is herbivores which eat plants and algae which are called primary consumers.
- The third trophic level is carnivores that eat herbivores and are called secondary consumers.
- The fourth trophic level is carnivores that eat other carnivores and are called tertiary consumers.

There may be additional levels beyond this but most food chains do not exceed six trophic levels. Apex predators are carnivores at the top of food chains that have no predators.

Decomposers break down dead plant and animal matter by secreting enzymes into the environment. These digest matter before it is absorbed into the decomposer. It is sometimes difficult to place decomposers into food chains, although they play a very important role in recycling nutrients in an ecosystem.

> **Exam tip**
>
> You should be able to describe the differences between the trophic levels within an ecosystem.

Pyramids of biomass   REVISED ☐

Biomass is living tissue and is measured in grams or kilograms. When we measure the biomass of organisms in the different trophic levels of an ecosystem we see the same pattern. This is called a **pyramid of biomass**. In a healthy food chain, the first trophic level will always have the largest mass. This is placed at the bottom of a pyramid of biomass. At each higher level, the mass decreases forming the pyramid. Because the mass of an organism can vary depending significantly upon on the water it has recently consumed, scientists usually measure biomass in dead, dried organisms.

> **Biomass:** Tissue from living or recently dead organisms.
>
> **Pyramid of biomass:** A graphical way of representing the mass of organisms at each level of a feeding relationship.

Typical mistake

Unless an ecosystem is in serious trouble, pyramids of biomass are always perfect pyramids.

Transfer of biomass

Photosynthetic plants and algae are only able to absorb 1% of the energy transferred by the Sun. Even this small amount is still enough to power most food chains on our planet. These producers are often eaten by primary consumers. They are only able to use ten percent of the total energy from the producers. This is converted into body tissues, so we can think of energy and biomass as almost interchangeable. A small amount of biomass is lost because:

- Not all the organisms at each trophic level can be absorbed into the next trophic level.
- Some is lost as faeces.
- Some absorbed material is then lost as waste (carbon dioxide and water in respiration and water and urea in urine).

The remaining energy is used by the producers for the seven life processes. Of these, a large amount of glucose is used in respiration.

Exam tip

You should be able to describe how biomass is lost between different trophic levels of a food chain. You should also be able to calculate the percentage efficiency of biomass transfer between trophic levels. Finally, you should be able to explain why the number of organisms often reduces at higher trophic levels.

Now test yourself

31 What do the arrows in a food chain show?
32 What types of organisms are found as producers at the bottom of most food chains?
33 What term is given to all organisms from the second trophic level and above?
34 How do decomposers feed?
35 What term do we give to organisms at the highest trophic level?
36 Define the term biomass.
37 What name do we give to a graphical way of representing the mass of organisms at each level of a feeding relationship?
38 What is true about all pyramids of biomass from healthy ecosystems?
39 Why are organisms killed and dried before being measured for biomass?
40 How is biomass lost from a food chain?

Answers on p. 117

Food production

Factors affecting food security

Food security is having enough food to feed a population. This situation is taken for granted in many areas, but tragically there are many people who are unable to earn enough money to feed their families and areas where food is so short that even those with money cannot. Biological factors that affect food security include:

- an increasing birth rate
- changing diets in developed countries means scarce food resources are transported around the world
- new pests and pathogens reduce farming yields
- environmental changes such as widespread famine occur in some countries if rain fails
- increasing costs of farming equipment and other inputs
- armed conflicts that affect the availability of water or food.

Farming techniques

REVISED

Humans began to evolve from hunter–gatherers to more settled farming communities around 12000 years ago. For many years these small groups produced the resources needed for their community to survive. Now more farms are owned by companies that have more pressure to earn money. This situation often creates conflict between ever-increasing yields and reducing damage to the local environment.

Monoculture or crop rotation

Monoculture occurs when farmers choose to grow only one crop. This allows the farm to specialise in one product, meaning it can be more effective in planting and harvesting. This can very quickly lead to nutrient deficiencies in the soil. This means that farmers either need new land or to add expensive fertilisers to their soils.

Crop rotation occurs when farmers grow different crops in the same fields in different years. This gives the soil time to recover and means that nutrient deficiencies do not occur. This is much more **sustainable** and less damaging method of farming.

Intensive farming

Intensive farming maximises farming yields, which creates the largest financial profits for farms. The following are examples of intensive farming:

- Fields are not left fallow to recover but are quickly planted again after the last crop was harvested.
- Larger numbers of animals are raised in smaller spaces which reduces their energy loss to the environment (**factory farming**).
- Fish and other animals are kept in cages to allow easy feeding of high protein foods to increase growth.
- Chemical fertilisers and pesticides, antibiotics and machines are used.
- Hedgerows are removed to create larger, easier to manage fields.

Organic farming

In many ways, **organic farming** is the opposite to intensive farming:

- Only natural fertilisers and pesticides are used.
- Crops are rotated and fields lie fallow to recover between crops.
- Animals are kept free range.

This costs the farmer more and so organic food costs the consumer more than intensively farmed equivalent food. Many people are now happier to pay for organic food knowing the way in which it has been produced is much less damaging for the environment.

> **Monoculture**: Sustained growth of one species of crop.
>
> **Sustainable**: Describes an activity that can continue without damaging the environment.
>
> **Factory farming**: Rearing livestock using highly intensive methods.
>
> **Organic farming**: Non-intensive farming that uses natural fertilisers and pesticides.

> **Revision activity**
>
> Compare and contrast the advantages and disadvantages of intensive and organic farming.

Sustainable fisheries

For many years we thought the stocks of fish in the oceans were limitless. We now think that over 85% of the world's fisheries have been overfished. A well-known example of a species that has been overfished is the North Sea cod. Between 1950 and 1980 the mass of cod taken from the North Sea increased nearly three times. We now understand better that we cannot take limitless amounts of fish without consequences. It is vital that we maintain fish stocks at a level at which breeding continues or certain species may disappear altogether in some areas. Since 1980 we have reduced the number of cod removed but not to below 1950 levels.

We have tried to reduce the number of cod being caught by:
- reducing the times boats can fish
- introducing **quotas** which limit the amount of fish caught
- fishing with nets with larger holes to only kill older fish that have reproduced.

This is a controversial issue. There is conflict between local fishing communities who depend upon this industry and conservation campaigners.

> **Quota**: A fixed entitlement to catch a specified volume of fish or other yield.

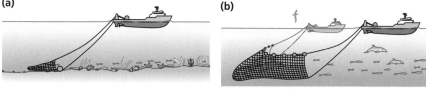

(a)   **(b)**

Figure 7.7 (a) Bottom trawling for plaice. What damage do the rollers do to the ecosystem? (b) Trawling for cod. What else gets caught in the nets?

Role of biotechnology

Biotechnology is the use of living organisms to develop and make products. This has been happening for thousands of years in agriculture and medicine. Recently our understanding of genetics has allowed us to genetically engineer new organisms and clone desirable individuals within species.

Biotechnological agriculture: mycoprotein

Protein that comes from fungi is called **mycoprotein**. This protein is suitable for vegetarians and has low levels of fat and high levels of fibre. A well-known brand of mycoprotein in the UK is called Quorn™. The fungus *Fusarium* produces mycoprotein. It is grown in large vats called **fermenters**. The fungus is grown on glucose syrup in aerobic conditions inside the fermenter.

> **Biotechnology**: The use of living organisms to develop and make products.
>
> **Mycoprotein**: Protein that comes from fungi, such as Quorn™.
>
> **Fermenters (vats)**: Giant containers with regulated conditions to maximise the growth of micro-organisms.

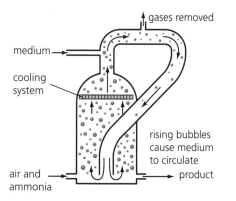

Figure 7.8 A fermenter for the production of mycoprotein.

Biotechnological agriculture: genetically engineered insulin

We have genetically engineered bacteria to produce human insulin. Before we could do this, diabetics injected pig insulin. Some diabetics were allergic to this, and others disagreed with it because they were vegetarians or for religious reasons.

Revision activity

Write out the procedure for genetic engineering as a flow diagram to help you remember.

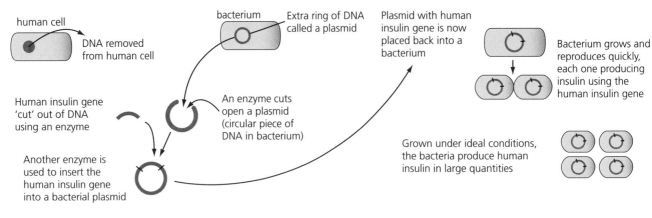

human cell

DNA removed from human cell

Human insulin gene 'cut' out of DNA using an enzyme

Another enzyme is used to insert the human insulin gene into a bacterial plasmid

bacterium

Extra ring of DNA called a plasmid

An enzyme cuts open a plasmid (circular piece of DNA in bacterium)

Plasmid with human insulin gene is now placed back into a bacterium

Bacterium grows and reproduces quickly, each one producing insulin using the human insulin gene

Grown under ideal conditions, the bacteria produce human insulin in large quantities

Figure 7.9 Genetic engineering of bacteria to produce human insulin.

Now test yourself

TESTED ☐

41 Describe some biological factors that affect food security.
42 What is monoculture and what are its consequences?
43 Define the term sustainable.
44 Describe some examples of intensive farming.
45 Describe some examples of organic farming.
46 What steps have we taken to prevent overfishing of cod?
47 What are quotas?
48 Define the term biotechnology.
49 What advantages does mycoprotein have over protein from farm animals?
50 What are fermenters?

Answers on p. 117

Summary

- A population is the total number of all the organisms of the same species in a geographical area. A community is a group of two or more populations of different species in the same geographical area. An ecosystem is the interaction between a community of living organisms and the non-living parts of their environment.
- In all communities, populations of organisms compete for resources. Plants compete for light, space, water and nutrients from the soil. Animals compete for food, mates and territory.
- Interdependence means all organisms within a community depend upon each other.
- Abiotic factors are non-living and include light intensity, temperature, moisture levels, soil

- pH and mineral content, wind intensity and direction, carbon dioxide levels for plants and oxygen levels for aquatic animals.
- Biotic factors are living and include availability of food, new predators arriving, new pathogens and one species outcompeting another.
- Both abiotic and biotic factors affect the distribution of organisms in an ecosystem.
- Adaptations are features which allow organisms to survive in the conditions in which they normally live. These can be structural, behavioural or physiological.
- Extreme environments include areas with high temperatures, pressures or salt concentration. The polar regions and deep-sea hydrothermal

- vents are examples. Organisms that live in these conditions are called extremophiles.
- Photosynthesising plants and algae are the producers of nearly all biomass (living tissue) on Earth. Food chains and webs show the feeding relationships of organisms within a community. A producer (usually a plant or alga) is found at the first trophic level. Producers are eaten by primary consumers (herbivores) which are in turn eaten by secondary and tertiary consumers (carnivores).
- The numbers and distribution of organisms in an ecosystem can be investigated by using quadrats and transects (imaginary lines along which quadrats are placed).
- Many different materials, including water and carbon, cycle through abiotic and biotic parts of an ecosystem. The carbon and water cycles are important for all living organisms. The carbon cycle returns carbon from organisms to the atmosphere as carbon dioxide to be used by plants for photosynthesis. Micro-organisms cycle materials through an ecosystem by returning carbon to the atmosphere and mineral ions to the soil. The water cycle provides fresh water for plants and animals on land before draining into the sea.
- Decomposition is the decay of waste material (rotting). The rate of decay is affected by temperature, water and availability of oxygen. Gardeners and farmers try to provide optimum conditions for decay of waste material into compost which is used as a natural fertiliser.
- Anaerobic decay produces methane gas which can be used as a fuel.
- Environmental changes in temperature, availability of water and the make-up of atmospheric gases affects the distribution of organisms in an ecosystem.
- Biodiversity is the variety of all the different species of organisms on Earth or within an ecosystem. High levels of biodiversity ensure stable ecosystems. The future of humanity on Earth probably depends upon us maintaining high biodiversity.
- A rapid recent increase in the human population and an increase in the standard of living mean we are producing more waste. If not treated appropriately, waste becomes pollution. This pollution can occur in water by sewage, fertilisers or toxic chemicals. It can occur in air from smoke and acidic gases. It can also occur on land from landfill and from toxic chemicals. Pollution reduces biodiversity.
- Building, quarrying, farming and dumping waste in landfill reduce the land available for other species. The destruction of peat bogs further reduces biodiversity.
- Deforestation has occurred to provide land for cattle and rice fields and to grow crops for biofuels.
- Global warming is caused by the greenhouse effect. Carbon dioxide and methane are greenhouse gases.
- Biodiversity is maintained by breeding programmes, protection and regeneration of rare habitats, reintroducing hedges, the reduction of deforestation and carbon dioxide emissions, and recycling.
- Trophic levels in a food chain are given numbers. The first is the lowest which is usually a plant or alga which makes its own food by photosynthesis. These are called producers. Producers are eaten by primary consumers (herbivores) which are in turn eaten by secondary and tertiary consumers (carnivores). Apex predators are found at the top of food chains. Decomposers break down dead plant and animal matter (biomass).
- The biomass of organisms at each tropic level can be shown in a pyramid of biomass.
- About 1% of incident light energy is used by photosynthesising organisms. About 10% of energy is transferred between trophic levels. The remaining is used by the organisms to complete the seven life processes. Because most energy is used, the number of organisms at higher trophic levels is often small and food chains usually do not exceed six trophic levels.
- Food security is having enough food to feed a population. Biological factors that affect food security include an increasing birth rate, changing diets in developed countries, new pests and pathogens, environmental changes, increasing costs of farming equipment and armed conflicts.
- Food can be produced more quickly by limiting energy loss in livestock. This can be achieved by limiting their movement and keeping them warm. Many are fed high protein diets. Intensive farming maximises farming yields by replanting fields quickly, keeping animals in cages or pens, the use of chemical fertilisers and pesticides, antibiotics and machines, and the removal of hedgerows.
- Organic farming is non-intensive farming that uses natural fertilisers and pesticides.
- The world's fish stocks are declining because of overfishing. The control of net sizes and quotas are important ways to preserve fish stocks.
- Biotechnology is the use of living organisms to develop and make products. An example is mycoprotein (protein from fungi) which is grown in fermenters. The genetic engineering of bacteria to produce human insulin is another example of biotechnology.

Exam practice

1 Which of these is not an abiotic factor? [1]
   A Light intensity
   B Temperature
   C Water availability
   D Disease

2 What are transects? [1]
   A Square frames of wire used for sampling
   B Imaginary lines along which sampling occurs
   C Small containers used to suck up insects
   D Large nets used to sweep though plants to collect insects

3 Define the term community. [1]

4 Describe the conditions near a deep-sea hydrothermal vent. [2]

5 Many young mammals can stand and run within hours of
   being born. Explain why this is an advantage. [2]

6 Describe and explain the predator–prey cycling shown for
   hares and lynx. [6]

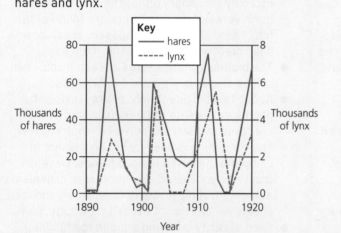

7 Describe the method you would use to investigate the effect
   of temperature on the rate of decay of fresh milk by measuring
   pH change. [6]

8 Describe the processes in the water cycle. [6]

9 Describe the method you would use see if there are more
   species of plant on the school field where it is cut or left
   uncut. [6]

Answers and quick quizzes online

ONLINE

29 How safe the supply of our food is.

30 Industrial agriculture to maximise yield, often involving the use of machines, chemical fertilisers and pesticides

31 The transfer of energy

32 Plants and algae

33 Consumers

34 By secreting enzymes into the environment and absorbing the broken-down products

35 Apex predators

36 Tissue from living or recently dead organisms

37 A pyramid of biomass

38 They are perfect pyramids.

39 Because the amount of water in their tissues can change their mass significantly.

40 Not all the organisms at each trophic level can be absorbed into the next trophic level. Some biomass is lost as faeces. Some absorbed material is then lost as waste (carbon dioxide and water in respiration and water and urea in urine).

41 An increasing birth rate, changing diets in developed countries, new pests and pathogens, environmental changes, increasing costs of farming equipment and other inputs and armed conflict

42 Sustained growth of one species of crop which leads to nutrient deficiencies

43 An activity that can continue without damaging the environment

44 Fields are not left fallow to recover but are quickly planted again after the last crop was harvested. Larger numbers of animals are raised in smaller spaces (factory farming). Fish and other animals are kept in cages to allow easy feeding of high protein foods to increase growth. Use of chemical fertilisers and pesticides, antibiotics and machines. The removal of hedgerows to create larger, easier to manage fields.

45 Only natural fertilisers and pesticides are used. Crops are rotated and fields lie fallow to recover between crops. Animals are kept free range.

46 Reducing the times boats can fish. Introducing quotas which limit the amount of fish caught. Fishing with nets with larger holes to only kill older fish that have reproduced.

47 A fixed entitlement to catch a specified volume of fish or other yield

48 The use of living organisms to develop and make products

49 It is lower in fat and has high levels of fibre.

50 Giant containers with regulated conditions to maximise the growth of micro-organisms.